WHY OBAMACARE IS WRONG FOR AMERICA

WHY OBAMACARE IS WRONG FOR AMERICA

How the New Health Care Law Drives Up
Costs, Puts Government in Charge of
Your Decisions, and Threatens Your
Constitutional Rights

GRACE-MARIE TURNER, JAMES C. CAPRETTA,
THOMAS P. MILLER, ROBERT E. MOFFIT

BROADSIDE
An Imprint of HarperCollinsPublishers
www.broadsidebooks.net

HarperCollins books may be purchased for educational, business, or sales promotional use. For information, please write: Special Markets Department, HarperCollins Publishers, 10 East 53rd Street, New York, NY 10022.

FIRST EDITION

Library of Congress Cataloging-in-Publication Data has been applied for.

ISBN: 978-0-06-207601-4

11 12 13 14 15   OV/RRD   10 9 8 7 6 5 4 3 2 1

Contents

Foreword

This important book provides an engaged citizenry with a clear diagnosis of why ObamaCare is the wrong prescription for American health care and points the way to genuine, patient-centered reform.

Despite all of the talk of ObamaCare's benefits, this book explains that the new law—misleadingly titled the Patient Protection and Affordable Care Act—will actually compound the worst problems in today's arrangements, most especially by driving out-of-control costs even higher.

Fortunately, while ObamaCare has been enacted into law, it is far from settled policy. A clear majority of the electorate favors its repeal.

And for good reason: as this book explains, this new law starts from the premise that the nation's health care system can, and should, be run from Washington, D.C. Under ObamaCare, power is shifted from patients, doctors, businesses, and states to the federal government. That will erode the doctor-patient relationship, lead to waiting lists for treatment, and foster widespread dependency on government-run health care. It will also impose a web of new regulations on businesses, stifling jobs and paychecks. On top of all this, it will add trillions in federal spending and push the nation more rapidly toward a debt crisis.

The result is that ObamaCare will make the health care cost problem worse, not better.

The exploding cost of health care is bankrupting the country, yet the government's open-ended commitment to subsidizing health care is the very reason costs are skyrocketing. Simply put, everyone who participates in the health care system is a rational economic actor, and ObamaCare will aggravate distortions in the current system and create perverse incentives for Americans to maximize their share of an ever-growing, increasingly opaque, and apparently limitless subsidy.

The authors of this book effectively contrast the two divergent paths for health care in America: the government-centric approach versus the patient-directed model. Decades of Medicare price controls have failed to control spending. Instead of cutting costs, price controls encourage providers to bill for more services each year. Medicaid is blowing out state budgets—there is no limit on the federal government's matching contributions to state spending, so state governments "spend most of their energy devising ways to 'maximize' how much they get from the federal government," while seeking to minimize their own share of the tab.[1]

The federal government also exempts employer-sponsored health insurance from taxation for workers. This policy tilts the compensation scale toward ever-greater (tax-free) benefits and away from higher (taxable) wages. This isn't just a big driver of runaway health care costs, as more dollars chase the same amount of services. It's also a big reason why too many Americans haven't seen a raise in a long time—the rising cost of health care is eating away at their wages.

The new health care law—with its maze of mandates, dictates, controls, tax hikes, and subsidies—will push costs further in the wrong direction. Already, health insurance companies have announced big premium hikes related to the law's new mandates. The law's so-called cost controls amount to the same kind of reimbursement cram-downs that have failed to control costs in Medicare. It will dramatically expand a Medicaid program that is already breaking state budgets and adding to a growing flood of red ink at the federal level. And it will create a new open-ended entitlement program at a time when we can't afford the entitlements we have.

We cannot find a bipartisan way forward on health care until Congress removes this partisan roadblock. While it is a prerequisite to real reform, repeal of ObamaCare alone will not be enough. We must also fix what's broken in health care, without breaking what's working.

In addition to clearly describing the problem, this book puts forward a framework for reform that will secure affordable, high-quality health care for Americans. It clearly illustrates how we must transform our government's reliance on defined benefits, which provide a false sense of security, to one of defined contributions, which provide the real security that comes with ownership, control, and flexibility.

This path to true choice and competition is reflected in health care reform proposals I've put forward in Congress. I'm proud to have worked with former Clinton administration budget director Alice M. Rivlin to advance some of these reforms, which is an indication that it is possible to build bipartisan support for policies that allow consumers and patients to make choices for themselves, even as the government provides sensible oversight of the marketplace.

In health care, as in any other economic arrangement, control of money is power. The question remaining is then: who gets the power, the government or the patient? Patient power will always serve the needs of the people far better than bureaucrats managing the decline of a government-run system on the verge of bankruptcy.

Giving patients and consumers control over health care resources would make all Americans less dependent on big corporations and big government for our health security, give us more control over the care we get, and force health care providers to compete for our business.

Over the next few years, the debate over ObamaCare will be critical to the health care all Americans receive and to the sustainability of the U.S. economy. The American people will play an indispensable role in this debate. Together, we can repeal this costly mistake before it hits with full force, and we can chart a new direction with true patient-centered health care reform.

Rep. Paul Ryan, Member of Congress

WHY
OBAMACARE IS
WRONG FOR
AMERICA

THE REAL STORY OF OBAMACARE

Here's the real story of how ObamaCare will impact you, why it cannot stand, and what we can do to put doctors and patients in charge of health care.

I n early 2009, as opposition to his sweeping health overhaul plan gathered strength, President Barack Obama tried to reassure his Democratic supporters that once the legislation passed, the poll numbers would quickly turn around. He told them privately, "I can sell this."

He was wrong. Never very popular to begin with, ObamaCare continues to face strong opposition from millions of Americans who are indignant that Congress passed it over their strong and vocal opposition. They know it is wrong for America. ObamaCare is wrong for families, wrong for patients, wrong for business, and wrong for our children's futures. The federal government will tell us that we must purchase health insurance and what it must cover, how much we can afford to pay for it, and how much we will be fined if we don't follow Washington's rules. It will take half a trillion dollars out of Medicare to create massive new entitlement programs we can't afford. And it will increase taxes by half a trillion dollars and cause health costs to go even higher.

How could this happen in America?

"Only in Washington, D.C., could you say you're going to spend a trillion dollars and save the taxpayers money.[1]

—Representative Mike Pence (R-Ind.)

THE PEOPLE RISE UP

The ObamaCare debate lit a fire under millions of Americans to become politically active, many for the first time. They rightly saw their freedom at risk.

Thousands of citizens had marched to the U.S. Capitol on a fateful Sunday in March of 2010, hoping one last time to stop passage of the massive health overhaul bill. Then-speaker Nancy Pelosi passed the demonstrators that afternoon as she and her entourage of committee chairmen linked arms and marched from the House office building to the Capitol across the street. She carried an oversize gavel to signal her determination and confidence that the bill would pass.

Police ringed the leaders as they walked past citizens carrying American flags and handwritten signs. "The people said No!" one read. But the citizens were ignored and derided—just as they had been over the many months leading up to the vote.

Some of the demonstrators who had stayed on into the night that Sunday to await the final vote gathered below the Members-only balcony on the south side of the Capitol building. The demonstrators arranged themselves on the lawn to spell the word "NO" to make it absolutely, crystal clear to anyone who ventured out where they stood. They were cheerful and hopeful to the end, believing that a few undecided members of Congress ultimately would listen to them and turn against the bill.

These were concerned citizens standing up for what they believe is right for America and its future. And their voices are still being heard.

THE 2010 ELECTION AFTERMATH

In the days leading up to the vote, the switchboard at the U.S. Capitol had been in near meltdown as 100,000 calls an hour came in from citizens across the country—the overwhelming majority pleading with their elected representatives to vote against the bill. One member of Congress took calls himself for an hour and said that not a single person asked him to vote for ObamaCare.

But President Obama, then-speaker Pelosi, and liberals in Congress were determined to ram their bill through no matter how strong the opposition. Many Democrats believed that passing health reform would be their permanent ticket to political power.

Americans initially were discouraged and demoralized after trying so hard to get Washington to listen to them and constantly being ignored and rebuked. But on November 2, 2010, voters got their chance to hold those who passed ObamaCare accountable for what they had done: any member who had voted for Obama-Care and who was in a competitive race was more often than not defeated for reelection.

A vote for the president's legacy health overhaul law was an albatross around the necks of candidates. One Democratic pollster, Pat Caddell, admitted that the health care legislation was the key reason for his party's loss of sixty-three seats in the U.S. House of Representatives.

Caddell told FOX News, "Among independents who favored repeal, 86 to 9 voted Republican. You could see Democrats going down who voted for health care—the health care bill—being wiped out . . . the American people found this a crime against democracy . . . they want it repealed, and this issue is gonna go on and on."[2]

Crucial independent voters were decisive. One Republican pollster, Bill McInturff of Public Opinion Strategies, found that 52 percent of independents said their vote in the midterm elections was a message opposing President Obama's health care law, and only 18 percent said their vote was a message of support. Those numbers were crucial in eroding support for Democratic House candidates in dozens of swing districts.

But President Obama and Democrats in Congress tried to discount the impact that ObamaCare had on the elections. The president insisted in his news conference the day after the midterm elections that the problem wasn't the policy but only that he hadn't explained it well enough. He said he would hold firm, adding that it would be "misreading the election if we thought that the American people want to see us for the next two years relitigate arguments that we had over the last two years."

He continued, "We can tweak and make improvements on the progress that we've made," but the only change he wanted was to put the overhaul plan in place even faster! Nancy Pelosi later said she had "no regrets" about her role in pushing the legislation into law.

Both are in denial that ObamaCare led to a historic defeat for Democrats. They still believe it is a communications problem and that people don't like the law because they don't know what's in it. So they are just going to try harder to explain it to us!

A CASCADE OF CASUALTIES

The Obama administration is refusing to accept the disastrous dynamics it has set into motion in the health care sector and economy. The American people know very well what's in the law and are already seeing the impact it is having, even before most of it takes effect. Here are a few of the early problems, with more detail throughout the book.

The bad news began just a few days after ObamaCare was signed into law. AT&T, Caterpillar, John Deere, Verizon, and several other big companies reported to investors—as they are required to do—that the new law would take a bite out of their future earnings. They were about to be hauled before the House Energy and Commerce Committee to explain their disloyalty until it became clear that they were likely to testify that they also are considering the option of dropping employee health insurance.[3]

Under the new law, companies will be fined $2,000 per employee if they don't offer health insurance, much less than the $7,000 or more that it costs per worker to provide comprehensive insurance. You can't blame them for working the numbers and considering dropping coverage and sending many of their employees into the subsidized state health exchanges for insurance.

Former Congressional Budget Office director Douglas Holtz-Eakin and our coauthor Jim Capretta write that employers will have strong incentives to move as many as 35 million workers out of employer plans and into subsidized coverage. They estimate that this would add about $1 trillion *more* to the total cost of Obama-Care over the next ten years, leading to a fiscal train wreck.[4]

The problem isn't just the cost. Businesses, both big and small, are aghast at the mandate to provide health insurance for their workers or face federal penalties. The threat alone is dampening job creation, especially with small businesses, which are usually the creators of virtually all new jobs in a recovering economy.

A fledgling insurance company in Virginia called nHealth was a very early casualty of ObamaCare. It shut down after investors concluded that it wasn't possible to navigate the maze of the new

ObamaCare regulations and still succeed. Other companies have announced that they are going to stop selling health insurance altogether.

Health costs are already rising, at least partly because of some of ObamaCare's provisions. Health and Human Services Secretary Kathleen Sebelius railed at insurance companies for explaining that an early wave of insurance mandates will actually increase premium costs. Instead of admitting the facts, the administration continues to lash out at people who have the audacity to tell the truth. "There will be zero tolerance for this type of misinformation and unjustified rate increases," she wrote in a letter to insurers.[5]

It's also already affecting hospitals. In late December, construction was halted on forty-five new physician-owned hospitals that were shut down directly as a result of the new law. If they weren't completed and certified by December 31, 2010, they would not be able to bill Medicare. These privately owned hospitals lost out to powerful hospital groups trying to eliminate the competition. Dr. Michael Russell, the president of Physician Hospitals of America, which has filed a lawsuit to try to stop the selective building ban from going into effect, says, "There are so many regulations and they are so onerous and intrusive that we believe that [the law] was deliberately designed so no physician owned hospital could successfully comply."[6]

The administration already is doling out waivers from its own regulations. For example, last fall, a number of companies with lower-wage employees said that the new regulations issued by the Obama administration would force them to drop health plans with limited benefit coverage because they couldn't meet the new requirements. Rather than face the prospect of hundreds of thousands of workers losing coverage, Health and Human Services

Secretary Kathleen Sebelius began granting one-year waivers to companies and unions that offer these so-called mini-med plans. By the end of the year, HHS had granted at least 733 waivers to companies such as McDonald's, Jack in the Box, and Ruby Tuesday and to a number of labor unions.[7]

The changes have only begun. Thousands and thousands of more pages of regulations will be written to put the legislation into force and will alter virtually every aspect of our health care sector.

Businesses went ballistic when they found a small provision in the law that will require them to file forms with the Internal Revenue Service (IRS) to report purchases from every vendor totaling more than $600 in a year. This has nothing to do with health care but was inserted in the bill at the last minute to raise $17 billion under the theory that more paperwork will catch tax cheats. The National Federation of Independent Business believes this barrage of paperwork will hit 40 million businesses.

Though the president repeatedly told Americans, "If you like your health care plan, you will be able to keep your health care plan,"[8] the facts show that won't be the case for tens of millions of working Americans and their families. A presentation by an analyst for the consulting firm McKinsey & Company showed that as many as 80 million to 100 million people could lose the coverage they have now and be switched into other policies or plans once the legislation takes full effect in 2014. Independent studies show that people with private insurance, people with insurance through their employers, seniors on Medicare Advantage plans, and children are all among those who may lose the coverage they have now.

Instead of acknowledging the consequences of what it has done, the White House rails against the companies and industries that are responding in perfectly rational ways to the incentive structures the health overhaul law has set up.

TRY IT, YOU'LL LIKE IT

President Obama insists he is "extraordinarily proud" of his signature legislative achievement and still believes that once we find out what is in the law, we will like it better.

Some of the opposition to ObamaCare has calmed because people are hearing mostly about a few provisions that took effect in 2010. The White House insisted on adding early sweeteners[9] to ObamaCare so it would have some benefits to talk about while the wheels of the regulatory machinery grind toward full implementation in 2014. The law allows, for example, twenty-six-year-old "children" to stay on their parents' health insurance and "free" preventive care with no deductibles or copayments, and it created new programs so some people with preexisting conditions can get insurance. But even these early changes are causing a backlash, with some people losing coverage and many others seeing the cost of their health insurance go even higher. For example, providing free preventive care simply means that costs are shifted to insurance premiums, making coverage more expensive. And the requirement that employers provide coverage for adult children caused some firms to simply drop coverage for all dependents—including a big labor union in New York that was a major supporter of passing the legislation.

There were better ways to solve these problems, and it could have been done with a few hundred pages of legislation, not a few thousand. But many people think that if this is all there really is to ObamaCare, what was the big deal?

The big wallop is coming when the full impact of ObamaCare

hits. It isn't clear what the president thinks people are going to like about what comes next. . .

The $575 billion that's going to be taken out of Medicare?[10]

Or the huge expansion of Medicaid that could bankrupt states or force them to slash spending on other services such as schools, police, and transportation?

The federal mandate that all of us must have expensive government-approved health insurance or pay a fine starting in 2014?

The $500 billion in new taxes on wages, investment income, medical devices, prescription drugs, and health insurance that will cause costs to go even higher?

Or maybe the many new, expensive rules and mandates on business that are discouraging the creation of new jobs?

Which part, exactly, are we supposed to like?

BIGGER BUREAUCRACY

The legislation itself runs to 2,801 pages when the two companion bills passed in the same week are combined.[11] But most of the major programs require an avalanche of regulations to explain in detail how they will work. These regulations will reach into every corner of our health care sector.

When the law creating Medicare and Medicaid was passed in 1965, it required only 137 pages of legislation. In 1998, Dr. Robert Waller, then the chairman of the Mayo Clinic, asked his staff to count the number of pages of regulations this premier health care clinic had to comply with to treat Medicare and Medicaid patients. The answer: More than 132,000 pages. So each page of legislation led to nearly 1,000 pages of regulations!

The expansion of bureaucracy alone is enough to give us ulcers. It's impossible to know how many bureaucrats will need to be hired at both the federal and state levels to run this cacophony of new programs under ObamaCare. It could run into the tens of thousands when the number of federal and state employees and contractors is included.

The law creates an estimated 159 new agencies, boards, commissions, and government offices dedicated to putting our health care system under Washington's control.[12] We say "estimated" because even the authoritative Congressional Research Service said it can't be sure how many new agencies ObamaCare will create.[13]

These bureaucrats aren't accountable to voters, and we will have virtually no control over the decisions they make on our behalf. In fact, we likely won't learn of their decisions until we find out too late that the government is cutting costs by altering how doctors take care of us. Two to watch are the Patient Centered Outcomes Research Institute, which will decide whether treatments are "effective" or not, and the Independent Payment Advisory Board, whose appointed officials will have sweeping powers over payments for Medicare treatments.

Many liberals believe that it's good to put all of those experts in charge. They say that health care is just too complicated for the average person to understand and we just aren't smart enough to make these decisions for ourselves. They believe they are doing us a service by making decisions for us—that they know best.

That kind of arrogance is exactly what has caused many Americans to rise up against ObamaCare.

A BETTER WAY

Nothing is more personal, or fraught with more consequences, than our health. Access to the right doctor and the right treatment at the right time can literally make a difference between life and death.

As we explain in later chapters, many doctors, especially those with small practices that can't withstand the avalanche of new government requirements and paperwork, are selling their practices or quitting medicine altogether. Some senior citizens are having trouble finding a doctor who will see them. The innovation that has drawn patients from around the planet to get their medical care here will be suffocated by taxes and regulation. The future under ObamaCare is bleak.

We needed health care reform, but not this! There are better ways to change policy, help people who need insurance, get care to those who are sick or hurt, and realign incentives to heal our health care sector.

Americans are a compassionate and generous people. One of the reasons we wanted health care reform is that we don't want millions of people to go without health insurance. Everyone should be able to get health insurance, but the changes shouldn't drive up the cost of insurance so millions more lose it.

But ObamaCare will leave at least 23 million people uninsured, spend well more than $1 trillion we don't have, lead tens of millions of people to lose the coverage they have now, and cause overall health care costs to rise even faster than if the legislation had not passed.

Worse, the changes will likely make it even more difficult for the most vulnerable Americans—people who need care the most and who can afford it the least. Dr. Edward Miller, the dean of the prestigious Johns Hopkins Medicine, worries that the huge expansion of Medicaid is going to make it even harder for institutions like his to serve "poor and disadvantaged" people who have nowhere else to go for medical care. He said, "the current health-care legislation will have catastrophic effects on those of us who provide society's health-care safety-net. In time, those effects will be felt by all of us."[14]

ObamaCare does not strengthen Medicare, as the president claimed. In fact, it targets seniors in the popular Medicare Advantage program, dumping many of them back into the fee-for-service market where care is often fragmented and uncoordinated.

Medicare is a government program, but one that is facing more than $30 trillion of debt. It's just not sustainable for the federal government to continue to write hundreds of millions of checks for hundreds of billions of dollars a year when there are 77 million baby boomers about to enroll in the program over the next two decades. The program must be reformed for the sake of today's seniors and future generations. Medicare savings should go into saving Medicare, not creating new budget-busting entitlements. If we start now, we can make changes that will be a win for taxpayers and seniors, as we will describe later in the book.

ARM YOURSELF WITH THE FACTS

To help win this fight, it's important to be armed with the facts about how the law is going to affect our families, our health care system, and our freedom.

Cynthia Tucker of *The Atlanta Journal-Constitution* wrote a column in January 2011 that's a good example of what we are up against.[15] Tucker praised PolitiFact—the fact-checking news bureau at the *St. Petersburg Times*—for its "Lie of the Year" award. The top prize went to the statement that the new health care law is a "government takeover of health care."

"Uttered by dozens of politicians and pundits, it played an important role in shaping public opinion about the health care plan and was a significant factor in the Democrats' shellacking in the November elections," Tucker wrote. "The lie was so successful that it helped to launch the tea party movement and the political victories of several of the new Republican members of Congress."

Tucker quotes PolitiFact's reasons for saying ObamaCare is *not* a government takeover:[16]

"Government takeover" conjures a European approach where the government owns the hospitals and the doctors are public employees. But the law Congress passed, parts of which have already gone into effect, relies largely on the free market:

- Employers will continue to provide health insurance to the majority of Americans through private insurance companies.
- Contrary to the claim, more people will get private health coverage. The law sets up "exchanges" where private insurers will compete to provide coverage to people who don't have it.
- The government will not seize control of hospitals or nationalize doctors.
- The law does not include the public option, a government-run insurance plan that would have competed with private insurers.
- The law gives tax credits to people who have difficulty affording insurance, so they can buy their coverage from private providers on the exchange. But here too, the approach relies on a free market with regulations, not socialized medicine.

SO HOW DO YOU ANSWER THIS?

The legislation that passed in March 2010 created the architectural drawings for the government-controlled system the administration is busily constructing. So if someone argues that ObamaCare

isn't a government takeover of health care, here are ten reasons
you can offer why it is:

1. For the first time in our nation's history, the government
   will order citizens to spend their private money on a private product—health insurance—and will penalize us if we
   refuse. U.S. District Judge Henry Hudson ruled in *Virginia
   v. Sebelius* in December 2010 that the individual mandate
   "would invite unbridled exercise of federal police powers"
   and that it "is neither within the letter nor the spirit of the
   Constitution." Another federal judge in Florida concurred,
   declaring the whole law unconstitutional.
2. The federal government will also determine what health
   care benefits are "essential"—not us, not our doctors, but
   government bureaucrats.
3. Doctors and hospitals will face an avalanche of new reporting rules to make sure they are providing medical services
   that fit the government's definition of "quality care."
4. The legislation creates a plethora of boards, commissions,
   and programs that will rule over virtually every corner of
   the health care sector and, for the first time, impose significant new federal regulations on private health care and
   health insurance.
5. States are treated like contractors to the federal government, not sovereign entities empowered by the Constitution. They are ordered to expand Medicaid to levels that
   could bankrupt them and to set up new health exchange
   bureaucracies lest the federal government sweep in and do
   it for them.
6. Any employer with more than fifty employees will be told
   it must provide government-decreed health insurance to its
   workers—or face financial penalties.

7. As many as 80 million to 100 million people will not have the option of keeping the coverage they have now, as President Obama promised.

8. As ObamaCare dramatically expands the number of people whose health coverage will be paid for entirely or in part by taxpayers, this expands further the government's power to decide which medical services millions more people will get—or not.

9. In order to parcel out taxpayer subsidies for insurance, the government and our employers are going to need to know a lot more about us. An estimated 16,500 more IRS agents will be needed to check on our income and any changes to our family status.

10. The law creates the infrastructure for public plans by requiring a federal agency to sponsor at least two national health plans. If private plans are crushed by ObamaCare's regulations or simply turn into government contractors, these government plans could dominate the market.

WHAT'S NEXT?

While ObamaCare is the law of the land and sinking its regulatory roots deeply into our health care sector, the new leadership in the U.S. House of Representatives is doing everything it can to slow it down.

Newly elected House Republican leaders put repealing and replacing ObamaCare at the top of the legislative calendar for the new Congress to "keep a promise to the American people." On January 19, 2011, the U.S. House of Representatives voted 245 to 189 to repeal the law, with every Republican and a few Democrats voting in favor of getting rid of ObamaCare, thus paving the way for subsequent legislation with more sensible reforms. More than two hundred economists sent a letter [17] to Congress the day of the vote saying that the president's health law is "a threat to U.S. businesses and will place a crushing debt burden on future generations of Americans."

But because Democrats still control the Senate, it's much more difficult to pass there. If the Senate were to break the logjam and pass the repeal bill, President Obama has vowed to veto it. So the main job of the 112th Congress will be to use the oversight and hearing process to make sure the American people are informed about the true impact and cost of the legislation, to conduct careful oversight on the regulations the administration is writing and the money it is spending to implement Obama-Care, and to pass targeted bills to dismantle, delay, and defund the law.

Meanwhile, the legal challenges to the law are moving forward, with at least two early court decisions declaring all or part of ObamaCare to be unconstitutional. The biggest legal challenges focus on the constitutionality of the federal requirement that all Americans must have health insurance and that states must vastly expand their Medicaid programs. The U.S. Supreme Court will surely take up the case, with a decision likely in the 2012 session. It's impossible to predict how that vote will go.

And there will be a lot of action with the states as they figure out how to deal with the avalanche of mandates. Many governors will be demanding much more flexibility, more time, and exit strategies to escape the bureaucratic and budgetary costs of ObamaCare.

If a new president were elected who supported repeal and if there were enough votes in Congress to get a repeal bill through both houses, the law could be wiped off the books. Then Congress could start over to develop sensible changes that fit with our economy and the will of the American people.

As the battles continue in the courts, in the states, in the Congress, in the media, and around dining room tables, all of us need to be armed with information for the battles ahead. This book will explain many of the things that are actually in the 2,801 pages of legislation. We'll explain what it will mean to you as families, young adults, senior citizens, people with health problems, physicians and other medical professionals, small-business owners and entrepreneurs, employers and employees, taxpayers, and citizens. You will see that there is some overlap of information in the chapters because most of us fit into more than one of these categories. We won't be able to cover everything, but we will tell you what we think is most important. We'll also explain what we think is a better path to health reform. And at the end of the book, we will offer suggestions about what you can do to put the brakes on ObamaCare.

WHY THE LAW WON'T STAND

The passage of ObamaCare was deeply polarizing. Never before had Congress passed—and the president signed into law—such sweeping legislation that was so strongly opposed by so many Americans.

ObamaCare is the central battle over the future direction of our country. It cannot stand. It is a looming disaster for patients— especially seniors—for our economy and for the future of health care in America. It will lead to rationing and government control over health care decisions. The quality of care will decline. Medical progress will slow. And the resulting third-rate health care system will likely bankrupt our country. The federal government is already racing toward a budgetary train wreck, and ObamaCare will only speed us down the tracks.

There is a lot of work ahead, but we can prevail in this fight to get to the right kind of reform. We don't trust government to make the key decisions about our health care, and we want our elected leaders to take out a clean sheet of paper and start over. We must replace this outrageous law with reform that will stabilize our health sector and our economy. The American innovative spirit is alive and well, and it can be brought into the health sector to give consumers more affordable health care options and newer and better options for health insurance.

Is there a silver lining to all of this? Some people think maybe so. ObamaCare has awakened a sleeping giant of Americans who had been hard at work with their jobs, raising families, and volunteering in their churches and communities, but who had been lulled into believing that our country would continue to prosper and that our freedom was sacrosanct. ObamaCare has awakened them to the dangers of giving the government too much control over our lives. If we can repeal this monstrosity, we will be on the road to securing our freedom and moving toward a more prosperous future where we are free, once again, to control our own destinies.

"America is not just a nationality . . . America is an idea. It's the most pro-human idea ever designed by mankind. Our founders got it right, when they wrote in the Declaration of Independence that our rights come from nature and nature's God—not from government." [18]

—Representative Paul Ryan

WHAT'S IN THIS LAW:
AN OVERVIEW

If I were designing a system from scratch, I would probably go ahead with a single-payer system.[1]

—*Then-senator Barack Obama*
August 18, 2008, Albuquerque, New Mexico

We have to pass the bill, so you can find out what's in it, away from the fog of controversy.[2]

—*Nancy Pelosi, then-speaker, U.S. House of*
Representatives, March 9, 2010, Washington, D.C.

After thirteen months of intense and bitter debate, marked by high drama and low parliamentary gamesmanship, President Obama finally did it. On March 23, 2010, he signed the Patient Protection and Affordable Care Act (PPACA) into law.[3]

It was also a power grab of the first order. In the teeth of popular opposition, Congress enacted the largest piece of social legislation in American history—and potentially the most consequential—and it did it on a purely partisan basis. Not one Republican in either house of Congress voted for the final law.

"You can't improve the health of a nation by bankrupting its children."[4]

—Representative Jeb Hensarling (R-Tex.)
House floor speech, March 21, 2010

President Obama's legislative victory had enormous political consequences. Congressional job approval sank like a stone— down to 18 percent by September 20, 2010.[5] The November 2010 midterm elections led to a historic defeat of Democrats who had supported ObamaCare. It's easy to see why. The new health care law takes giant steps toward putting government in charge of a huge swath of our economy. And that is exactly what the American people oppose.

We are already approaching a tipping point. Government actuaries say health spending in the United States will total $2.7 trillion in 2011—that's one out of every six dollars spent in our entire economy. And for the first time in our nation's history, more of the payments for health care will come from government than from the private sector.[6]

Nonetheless, members of Congress decided they knew better than we did what was good for us. They believed that government would be able to make countless decisions—big and small—about how this highly complex part of our economy will be organized and run. So they prescribed stronger doses of government-run health care, insisting on creating a big bureaucracy that will make key decisions, directly or indirectly impacting health care for all of us. Most congressmen and senators couldn't hope to read all 2,801 pages of the two bills that make up ObamaCare, let alone fully comprehend them—especially not within the tight deadlines set to get the bills passed.[7] This chapter will give you an overview of what's in the law—the things that members of Congress should have known before they voted for it.[8]

THE TOP TEN LIST

The scope of the new law is breathtaking. One new federal requirement leads to another, and another, and another. And the whole thing is paid for with an avalanche of new taxes and deep reductions in what Medicare will pay for services.

Explaining in detail everything in the law would take thousands of pages, but here is a list of the top ten things ObamaCare will do.

1. Most Americans will be required to buy federally approved health insurance starting in 2014, and the government will fine us if we don't comply.[9]
2. Companies will have to provide or pay for health insurance for their employees or pay a penalty.
3. The federal government will tell us what health benefits must be covered in these policies, as well as what share of our income we will have to spend on health insurance.
4. New federal entitlement programs will be created to provide taxpayer-subsidized benefits and insurance to tens of millions more people.
5. States will be required to set up new agencies and bureaucracies to restructure their health insurance markets, certify that health plans meet federal standards, and qualify people to receive taxpayer subsidies for health insurance.

6. States will be required to expand Medicaid—the program originally intended for the poor—to cover as many as 84 million people by the end of the decade, a mandate many states say could bankrupt them.

7. Dramatic reductions will be made in payments to Medicare providers and Medicare Advantage plans to partially pay for the expanded health coverage.

8. More than $500 billion will be raised in new taxes in the next ten years, most of which will be passed on from businesses to consumers.

9. A massive federal bureaucracy will be created, with an estimated 159 new federal agencies, boards, commissions, offices, panels, and spending and grant programs.[10] But since new offices also can be created by the administration without an act of Congress, the Congressional Research Service says that the exact number is "unknowable."

10. A "public option" will be created to compete against private health insurance—though administration officials don't call it that. The law requires the U.S. Office of Personnel Management to sponsor at least two health plans to compete nationally against various "local" private health plans in the state-based health exchanges. These plans will inevitably become the de facto "public options" that were supposedly left out of the bill.

THE ARCHITECTURE OF GOVERNMENT CONTROL

In a nutshell: ObamaCare will collect more than $500 billion in new taxes and take $575 billion from Medicare over the next ten years. Almost everyone will be required to have federally designed health insurance, and most employers will be required to provide insurance or face fines. A vast network of new bureaucracies will be created to reorganize the health insurance market and to develop and enforce all of the new federal rules. Two new entitlements, plus a big Medicaid expansion, are created to reduce the number of uninsured at a cost of at least $2.3 trillion over the first ten years of full implementation.

Many people say it is about time that the United States joined the rest of the civilized world in providing access to health insurance for all. Many people in other countries mistakenly believe that if people in the United States don't have insurance, they are denied access to health care altogether.

The U.S. system of providing care to the uninsured is indeed fragmented—from requirements that hospitals treat patients whether they can pay or not to charity care provided by private physicians, community health care centers, and free clinics, to people who pay for their own care out of pocket.

We can do better. But instead of developing a uniquely American solution which builds on the strengths of our economy and fixes the underlying causes of our problems, ObamaCare relies on many of the tools of European social welfare states.

The health overhaul law reflects an ideology that moves power and control away from individuals and toward government and its agents. This will inevitably lead to a society with a much larger role for government and greater government redistribution of wealth. If this transformation succeeds, the United States' welfare state would expand dramatically, and our nation would look much more like Western Europe.

An adviser to Presidents Obama and Bill Clinton actually said it best. Paul Starr, in his book *The Social Transformation of American Medicine*, wrote, "Political leaders since Bismarck seeking to strengthen the state or to advance their own or their party's interests have used insurance against the costs of sickness as a means of turning benevolence to power."[11]

ObamaCare's mandates and regulation of health insurance and expansion of government entitlement programs will dramatically change how we get health insurance and health care in the future. We will sacrifice a great deal of our freedom if Obama-Care is put into place.

OBAMACARE: WHAT'S IN THIS LAW?

The edifice of ObamaCare rests on eight pillars:

1. Federal mandates on individuals
2. Federal mandates on employers
3. Expanding federal entitlements
4. Squeezing funds out of Medicare and choking off private plan choices
5. New federal taxes
6. Health insurance exchanges
7. Federal government-sponsored health plans
8. Federal control over private health insurance

Pillar 1: *Federal Mandates on Individuals*

Under ObamaCare, you will be required to have federally approved health insurance. If you are on Medicare, Medicaid, on another federal program, or are in a military plan, your coverage is expected to satisfy the mandate to buy insurance. Almost everyone else will have to buy the health insurance that Washington approves or face fines, unless you can get a waiver from Washington because it's too expensive for you.

Beginning in 2014, you will face a penalty if you don't purchase a health policy that meets the federal government's definition of a "minimum essential" level of health coverage.[12] Those penalties will be enforced by the IRS. House Republican appropriators are trying to defund part of ObamaCare so the IRS doesn't have the funds to hire the additional 16,500 agents needed to enforce ObamaCare's rules.[13]

The penalty for not having health insurance will be either a flat dollar amount or a percentage of your income, whichever is greater. It is to be phased in over a period of three years and could be up to 2.5 percent of your income.[14] That means if your income is $100,000, you would pay a fine of $2,500 each year that you don't have health insurance.

But the penalty doesn't apply to everyone. If you are an illegal immigrant, a foreign national, a member of a Native American Indian tribe, qualify for a religious exemption, are in jail, or if the health insurance costs more than 8 percent of your household income, you are not subject to the penalty.

If you think you can start your own "religion" and get a religious exemption, you will be disappointed. You would have to prove that you're a member of a recognized religion with "established tenets or teachings" that would prohibit you from enrolling in health insurance.[15] The Amish and Mennonites, for example, would qualify. Another exemption is available to persons in "health-sharing ministries." These are nonprofit organizations made up of individuals who join together to finance one another's medical bills. It's not traditional insurance, but it will qualify for an exemption.

The HHS secretary can also grant a "hardship" exemption to people who can make the case that they can't afford insurance, even with the taxpayer subsidies and other exemptions. She would determine whether they qualify for such an exemption—after submitting the requisite paperwork. Otherwise, the IRS must enforce the individual mandate and collect the penalties.[16]

Pillar 2: *Federal Mandates on Employers*

Most Americans under age sixty-five get health insurance through their employer. Under the new law, your employer and federal regulators will make the crucial decisions about whether you can keep that coverage. If your employer decides to drop coverage, he or she could send you to the new state-run health insurance exchanges to get your insurance. This could easily happen to tens of millions of Americans.

If you are an employer, you will face some critical decisions before January 1, 2014. If you don't presently offer health insurance to your employees and you have more than fifty full-time employees,[17] you will be required to provide them with health insurance. And the insurance must meet the federal government's standards. Otherwise, you must pay a penalty of $2,000 a year for each of your full-time employees.[18] You will get a "free pass" from penalties on the first thirty employees. So if you have fifty-one employees and don't offer health insurance, you would pay a $2,000 fine on twenty-one employees. Your penalty payment would therefore be $42,000 a year. If you have one hundred employees, your fine would be $140,000—each and every year. It's no wonder that employers are hesitant to hire new workers with this job-destroying mandate hanging over their heads.

It gets worse. Let's say you have an employee whose family income is $88,200 a year or less (in 2010 dollars). If the insurance you offer is deemed "unaffordable" for him or her and your employee gets subsidized health insurance through the exchange, you would pay a penalty of $3,000 a year for that employee. This especially discourages employers from hiring lower-wage workers, who are certain to qualify for subsidies.[19]

It gets worse still. In order to avoid penalties like this, as an employer you must find out the household incomes of your employees, not just the wages you are paying them. That could be tricky. Household income would include the income of their spouse and working children living at home, as well as investments. As an employer, it is unlikely you will have a clue about that information. You may pay a worker the same wages and keep your health plan exactly the same. But if his or her household income changes, you could still face a fine because a spouse is laid off and your employee's household income falls. If you don't know about this, you could be hit with a $3,000 penalty—even when you didn't change a thing.

As an employer, you may consider another option. Beginning in 2014, you will likely have some employees who don't participate in your health plan because they say they can't afford it. In that case, you could be required to offer each of those employees a "free choice voucher" to buy their health insurance in the exchange. The amount of that voucher would have to equal what you would have spent to enroll the employee in your company plan. If you provide that voucher, you would not be subject to the $3,000 tax penalty.

If you are a large employer with two hundred or more full-time workers, you must automatically enroll your employees in your health plan.[20] They may choose to opt out and get their insurance somewhere else. And you must report which of your employees accept health insurance and which reject it, including whether they accepted it for themselves only or for their family.[21]

If you own a small business, you could be eligible for a tax credit. The credit could be worth 35 percent of the cost of insurance. But to get the full credit, you must have ten or fewer workers, pay an average annual wage for these workers of $25,000 or less, and pay at least half of insurance premium costs.[22] If you are self-employed, however, you cannot claim the small-business credit for insurance purchased for yourself or your family.[23]

"Our . . . recent health reform has created a situation where there are strong economic incentives for employers to drop health coverage altogether. The consequence will be to drive many more people than projected—and with them, much greater cost— into the reform's federally subsidized system."[24]

—Former Tennessee governor Philip Bredesen

Pillar 3: *Expanding Federal Entitlements*

The new law requires a major expansion of Medicaid—which is already the largest health coverage program in the country. It also creates two new federal spending programs: one to provide taxpayer subsidies to people who buy health insurance in the exchanges and another for government long-term care insurance, known as CLASS.

Medicaid Expansion. Beginning in 2014, the new law requires your state to expand Medicaid substantially. This is the program that was designed to provide safety-net coverage for the poor and indigent.[25] Under the new law, people under age sixty-five who have incomes below $14,400 for an individual and $29,300 for a family of four (133 percent of the "federal poverty level," in governmentese) would be eligible for Medicaid.[26]

Federal taxpayers will pick up 100 percent of the cost for this new Medicaid coverage until 2016, and then the federal matching payment begins to drop to 90 percent. Funny thing, though: Washington seems to forget that most people are both federal and state taxpayers. So there's no net new money here. They're just taking it out of a different pocket. Total federal and state Medicaid spending will skyrocket, going from $427 billion to $896 billion between 2010 and 2019.[27]

New Taxpayer Subsidies. Beginning in 2014, ObamaCare creates a new system of taxpayer subsidies for individuals and families to offset the cost of health insurance. The subsidies are available only if people buy their health insurance through the state-run exchanges. As we noted, the new law provides subsidies for families earning up to $88,200 (in 2010 dollars).

The subsidies will be generous, and they will vary by income. For example, if you make $59,250 annually, your family would get a premium subsidy of about $7,200. If your family income is $71,000, you would get a subsidy of about $5,200.[28] The Congressional Budget Office (CBO) has estimated that by 2016, an average health insurance policy for a family of four would cost approximately $14,100.[29] The taxpayer subsidies are deliberately structured to insulate individuals and families from the total cost of health care and insurance, but families would still face significant costs.[30,31] Still, it is a gigantic shell game. We all are paying, one way or the other—whether through higher taxes, deficit spending leading to higher taxes in the future, lower wages, or even lost jobs.

The CLASS Program. Congress also created a new government "insurance" program designed to help people who are disabled or elderly with the costs of long-term care. It's called the Community Living Assistance Services and Supports Act (CLASS). If your employer participates in the new program, you and all other working adults in the firm will automatically be enrolled through your job unless you opt out. If you don't opt out, a monthly premium will be automatically deducted from your paycheck. The estimated cost will be $146 a month, initially.[32] You must pay premiums for five years before you can expect any payments for benefits. As for the benefits, there are no upper limits. There is a floor, though: the law says they can't be less than $50 a day, adjusted for inflation.[33] Our chapter "Impact on . . . Taxpayers" explains why this is a Ponzi scheme and is one of the many ways ObamaCare's costs will be much higher than advertised.[34] In its initial assessment, the Office of the Actuary at the Centers for Medicare & Medicaid Services (CMS) predicted that the spending on CLASS will exceed the premiums, will be more attractive to older and sicker workers than younger and healthier ones, and will become financially "unsustainable."[35] There's a better way to provide long-term care assistance and to help people prepare financially for future needs without creating yet another unfunded government entitlement program.

"A Ponzi scheme of the first order, the kind of thing that Bernie Madoff would have been proud of."[36]

—Senator Kent Conrad (D-N.D.), referring
to the new CLASS entitlement program

Pillar 4: *Squeezing Funds out of Medicare and Choking off Private Plan Choices*

One of the ways that ObamaCare expands insurance coverage is through $575 billion in "savings" from Medicare over ten years.[37] Medicare already has financial problems, and there certainly are savings to be found. But these savings should go back into making Medicare stronger and more sustainable. Instead, the savings will go to fund huge new and expanded entitlement programs.

If you're a senior or a long-term disabled person, you, along with 47 million of your fellow citizens, almost certainly are enrolled in Medicare. Under the new law, you'll see big changes in health care delivery, including the biggest spending reductions in Medicare's history.

Here are some of the ways the new law will affect you:

First, if you're enrolled in a Medicare Advantage plan, the new law will reduce payments to your plan. In 2011, payments will be frozen, and more payment reductions will be phased in over the next few years. The CBO estimates that savings from cuts in payments to Medicare Advantage plans will total $136 billion over the first ten years. Many Medicare Advantage plans will be forced to exit the market, moving seniors back into traditional Medicare, where many will have to purchase separate supplemental coverage.

Second, payments to hospitals and other providers who treat Medicare patients, such as skilled nursing facilities, home health agencies, and hospices, will be ratcheted down. The law also changes payments for ambulance services, ambulatory surgical services, laboratory services, and some durable medical equipment. Fewer and fewer of these providers will be able to afford to treat Medicare patients. The Office of the Actuary at CMS says those payment changes will jeopardize seniors' access to care.

Third, Medicare will have other new rules designed to save money in the way health care is delivered. We do need incentives for doctors and hospitals to provide better care, not just more care. However, ObamaCare offers mostly an odd mixture of price controls, complex payment formulas, and new penalties for non-compliance. The law also provides for new organizations that are presumed to provide better coordinated care, called "Accountable Care Organizations." Doctors, hospitals, and other medical professionals are supposed to work together to take care of patients and share in any savings from delivering care more efficiently. (Read more about them in our chapter "Impact on . . . Seniors.")

The new rules will affect how your doctor practices medicine and how Medicare pays your doctor. If your doctor is a primary care physician, the new law will provide him or her with a 10 percent bonus, and the same will apply for surgeons in areas of the country that the federal government decides are "shortage" areas. For those doctors and surgeons, the bonus payments will be a welcome relief. Generally speaking, though, doctors will likely find practicing in Medicare even more onerous.

For example, the law will pay doctors according to how they comply with government standards for quality, and doctors must report required information to HHS.[38] If your doctor follows Washington's new rules, he will get a Medicare bonus payment; if he doesn't, his Medicare payment will be cut. By 2015, your doctor's participation in the quality-reporting program—assuming he wants to continue taking Medicare patients—will be mandatory.

The law does nothing to fix the flawed formula for how Medicare pays physicians, which threatens Medicare doctors with huge payment cuts each year.

As noted, the government expects to save $575 billion from these Medicare payment reductions and other program changes over the next ten years.[39]

A Powerful Board. The law also creates the Independent Payment Advisory Board (IPAB) to make recommendations to slow the growth of spending in Medicare. This board's job is to figure out how to reduce future Medicare spending with "specific and detailed" proposals. The board will have fifteen members appointed by the president and confirmed by the Senate. Its task: to bring the growth of Medicare spending into line initially with measures of inflation and later with measures of economic growth.

The only real tool it has is to recommend that providers get paid less or reduce payment for specific items or services. The board cannot legally recommend cutting benefits, changing cost-sharing, or adopting broader Medicare reforms. Hospitals will be exempt from the board's recommendations until 2019 (they are subject to other payment reductions before then, apart from the IPAB). Under the new law, the HHS secretary will implement the board's recommendations unless Congress passes an alternative proposal achieving the same level of Medicare savings.

If you are a senior citizen or baby boomer (or are about to become one), be aware that this is the very first time in the history of Medicare that Congress has enacted a hard cap on the program's spending. There will be consequences. And the members of the IPAB aren't elected and can't be easily fired if you don't like what they decide.

Pillar 5: *New Federal Taxes*

Approximately half of ObamaCare's costs will come from the cuts in payments to Medicare providers we described above. The other half will come from increasing your taxes. The tax increases in ObamaCare will total $503 billion between 2010 and 2019, based on estimates of the Joint Committee on Taxation.[40]

Some of those new taxes will have nothing to do with health

care. For example, there are tax increases on investments and on higher-income taxpayers. And if you run a business, you will be required to report to the IRS your purchases totaling more than $600 from any vendor in a year. This is supposed to close tax loopholes and raise an additional $17 billion over ten years. But Congress completely disregarded the mountain of paperwork it will impose on businesses, big and small. Not surprisingly, this provision of ObamaCare has become highly unpopular among voters and in Congress, even among House and Senate Democrats who voted for it. It is one of the first provisions of ObamaCare likely to be repealed before it takes effect in 2012.

Most of the new taxes in the Patient Protection and Affordable Care Act, however, are directly related to health care. They will affect almost all Americans directly or indirectly. They are not just taxes on "rich people."

The special tax deduction for Blue Cross/Blue Shield now is limited, so if you are enrolled in a Blue Cross/Blue Shield plan, that could affect your premiums. Likewise, if you use the services of an indoor tanning service, you will feel the effects of a 10 percent excise tax. Other new taxes will hit health care spending accounts that are presently tax-favored.[41]

Most of the major new taxes will come in 2013. If you and your spouse earn more than $250,000 a year (or you alone earn $200,000 or more), your Medicare payroll tax will increase from 2.9 to 3.8 percent. A new 3.8 percent tax will also apply to "unearned" or investment income, once you earn more than the previous income threshold. This is a huge change in tax policy. This 3.8 percent tax will target income from stocks, bonds, dividends, rents, and even, under certain circumstances, the sale of your home! CBO says that these new Medicare taxes will account for $210 billion in additional federal revenue over the period 2013 to 2019.

In 2013, there will be a new 2.3 percent tax on certain medical devices. If you depend on these devices, expect that tax to be passed on to you in the form of higher prices.

In 2014 and every year after that if you don't have a government-approved health insurance policy, you will have to pay a penalty. Whether Washington officials call it a penalty or a tax is ultimately irrelevant for budgetary purposes (but not for legal purposes, as we note in the chapter "Impact on . . . You and Your Constitutional Rights"). The bottom line is that it becomes revenue to the government to help finance ObamaCare. Think about it: tens of billions of dollars in financing for the new law depends in part on a sufficient number of persons remaining uninsured and paying a penalty for breaking the law. Likewise, your employer will have to pay a tax penalty if it doesn't offer the legally prescribed level of coverage to you and your fellow workers. By 2019, the government is expected to collect about $120 billion from those who don't comply with its mandates.[42]

Also in 2014, there will be a new tax on health insurance companies. That will surely be passed on to you in higher premiums. If, instead of relying on health insurance for routine health care bills, you would rather spend more out of your tax-free flexible spending account, in 2013 you will be faced with a new limit on what you can deposit in that account tax free: $2,500 annually.

In 2018, if you are still enrolled in what the Congress has defined as a "high-value" health insurance policy with "Cadillac coverage"—a plan costing at least $10,200 a year for individuals and $27,500 for families—your insurance company will pay an excise tax of 40 percent on the excess above that amount. This special tax is expected to yield $32 billion over the first two years of its enforcement.

Pillar 6: *Health Insurance Exchanges*

President Obama had to compromise somewhat on his goal of having one national exchange to regulate health insurance and distribute subsidies. Instead, he got fifty of them, but those exchanges will be "regulated" by HHS.[43] While each state will be required to set up and run its own exchange, the exchanges will still be the central vehicle for the federal government to control and regulate the health insurance market. Washington will dictate how the exchanges must work—and it will sweep in and set them up if the states don't do it themselves.[44] The exchanges have to be up and running in 2014.[45]

The federal government will require the exchanges to perform about a dozen "minimum functions."[46] They will have to check the incomes of people applying for health insurance subsidies to make sure they qualify—every month. They will have to make sure that the private plans offering health insurance meet federal standards and that they have enough and the right mix of doctors. Some of the functions will be purely administrative, such as ensuring that plans use a "uniform enrollment form" and a standard format to explain benefits. Each exchange also must have a website so people can compare plans on the Internet. But Washington is watching: even those administrative functions must follow the rules and guidelines established by the HHS secretary or other federal officials.

People who are eligible for federal subsidies for health insurance through the exchange must choose from one of several health plans that meet specific federal rules. What type of plan qualifies? One that meets the standards to be set by the HHS secretary. The secretary has sweeping powers to decide which insurers will be allowed to offer policies through the exchanges.[47] Health and Human Services Secretary Sebelius already has started threatening plans, telling them that if they don't play by her rules, she may shut them out of those markets. Few companies could afford to disregard this market, and therefore they have added incentives to play by Washington's rules.

There are some plans that the law says the secretary can't shut out. The law charges the federal government's Office of Personnel Management (OPM), which administers the health insurance program for members of Congress and others in the federal workforce, with creating new government-sponsored multistate health plans. The plans will be administered by OPM and "deemed" to be certified to offer policies in the exchanges.

In running the exchanges, your state officials are free to say that the health plans must provide *more* benefits than Washington says—thus making your health benefits package more expensive—as long as the state is willing and able to subsidize the extra costs this will create. But they cannot certify plans with fewer benefits. So there is little chance of making your health insurance less expensive.

Your state officials are also authorized (and strongly encouraged) to review your health plan's premiums. If they decide that the premiums are too high, they can exclude the plan from the exchange. Secretary Sebelius says that if any plan has premium increases of 10 percent or more, it will automatically trigger a review.

Though federal law requires the state-based health insurance exchanges to be set up in the states, they are subject to the HHS secretary's broad authority to issue rules and to set standards for them.[48] States must follow and enforce those federal rules and regulations. They aren't allowed to issue rules or regulations that block or conflict with federal rules and standards. If your state officials want to do something other than establishing the state-based health insurance exchange, they can do so only by getting permission from Washington.[49] And they have to wait until 2017 to ask for a waiver.

Pillar 7: *Federal Government–Sponsored Health Plans*

Beginning in 2014, as we noted, the law will require the federal government to sponsor at least two national health coverage plans. They will compete against private health insurance in the state-based exchanges. Under the law, at least one of those plans must be nonprofit, and at least one must not cover abortions. There may be more plans sponsored by OPM that could crowd out private plans trying to compete nationwide.

The deck is stacked in their favor. The Office of Personnel Management has broad authority to set the rules for the plans, and they can be less rigid than some of the most onerous rules that the private health plans must follow—such as how much they must spend on premiums versus administrative costs, their profit margins, and the premiums they can charge.[50] The government-sponsored plans will have to meet state licensure and other state health insurance requirements, but the scales will still be tipped in their favor.

The government-sponsored plans automatically qualify to offer policies in the exchanges. Obviously, big insurance companies will have the inside track to contract with the government to become one of the government-sponsored plans.[51] They will doubtlessly end up enrolling millions of your fellow citizens. Small and regional carriers will be at a serious disadvantage.

There is some controversy as to whether or not the new law contains a "public option," a key goal of the president and progressive policy analysts. The Kaiser Family Foundation, a prominent foundation that conducts extensive research on health care and health care policy, categorizes the OPM-administered plans under the heading "public plan option."[52] Dr. Stuart Butler, a distinguished fellow at The Heritage Foundation, observed that "in reality, this decision by Congress may actually open up the path to a much tougher public plan option than even House advocates imagined."[53]

ObamaCare also authorizes—and provides $6 billion—for the Consumer Operated and Oriented Plan (CO-OP) program. It will help set up nonprofit, member-run health insurance companies in all fifty states. These nonprofit companies will have to be self-governing, and all profits will have to go toward reducing premiums or enhancing benefits or the delivery of quality care.

Pillar 8: *Federal Control over Private Health Insurance*

The federal government will dictate the smallest details about your health insurance—even down to the wording in marketing brochures. Beginning in 2014, health insurance companies will have to follow federal rules in how they market their plans. No matter how complex the rules the government sets, all plans will be required to provide the information to you in plain language,[54] unlike the thousands and thousands of pages of rules and regulations the government regularly hands out.

The federal government will be comprehensively regulating health insurance benefits and markets, which have always been within the purview of the states. The new rules include: (1) "guaranteed issue," meaning that health insurance companies must sell you a policy even if you wait until you are sick to buy it; and (2) price limitations on how much more or less the insurer can charge you based upon your age. Health insurance for an older person can't cost more than three times what a younger person would pay for the same coverage, with a few exceptions.[55] Insurance companies cannot cancel coverage except in cases of fraud, and they cannot make a person wait any more than ninety days to enroll.

Second, the law also says that, starting in 2014, insurers can't put any annual caps on what they agree to pay out in claims.[56] That sounds good, but it means that a lot of people with less comprehensive insurance simply won't be able to afford their policies.

Third, beginning in 2014, the federal government will determine what must be covered by the health plans we all will be required to have. Once again, the HHS secretary will decide what are "essential health care benefits." She's going to give us the list sometime in 2011. Expect it to be very long and very comprehensive. Expect millions and millions of dollars to be spent by every health care company and provider who can afford to hire lobbyists to make sure their products and services make the list. And what do we expect this extremely comprehensive insurance to do to the price of the mandatory health insurance? You don't have to guess.

Fourth, ObamaCare sets strict limits on what our deductibles can be and what cost-sharing arrangements are allowed. It also establishes four types of health plans that can be offered through the exchanges, as well as the individual and small-group health insurance markets.[57]

Fifth, health plans will be subject to federal "medical loss ratio" (MLR) rules. This means that health plans in the individual and small-group markets must spend 80 percent of their revenues on medical benefits, retaining no more than 20 percent for the costs of administration, marketing, other customer services, and profits. For the large-group market, 85 percent of premiums must be spent on medical benefits. If the MLR targets are not met, the health plans will be required by law to refund the difference to enrollees in the form of rebates.

Finally, beginning in 2010, the new law provides for a process whereby states can review increases in health insurance premiums and determine whether they are justified.[58] If an insurer is found by government officials to have imposed "unjustified premium increases," the insurer must submit a written justification for them. The federal government will make grants to the states, totaling $250 million, over the next five years to implement this review process. This authority, as noted, will give government officials the power to exclude health plans with "unreasonable rates" from competition in the exchanges. In 2014, the HHS secretary will also monitor premium increases both inside and outside the health insurance exchanges.

POWER TO WASHINGTON

ObamaCare will affect far more than your health care. It will transform the way in which we are governed.

For those who designed ObamaCare and who wanted the federal government to control key decisions in the health care sector, the Patient Protection and Affordable Care Act is their path to success. The new law engineers a massive transfer of authority from citizens and the states to the federal government.

As we have described, there is an expectation by the law's authors that the states will dutifully implement the sweeping new provisions of ObamaCare. States are required to expand their Medicaid programs, which are already bankrupting many of them. They must set up massive new bureaucracies to deliver federal subsidies for the new highly regulated government-approved insurance plans. And they must stand by while the federal government micromanages their health insurance markets. Under the new law, the states will be subordinate to Washington; they will not be equals. American federalism, based on the equality of federal and state authority, each supreme within its own sphere, will be gravely injured.

The major changes in Medicare ahead will exacerbate access problems in health care delivery. The practice of medicine will simply not be the same. Moreover, the independence of physicians will be further weakened by having more and more patients enrolled in government health programs that pay less than private plans.

The language of the legislation, unfortunately, is an insufficient guide as to how exactly the new law will affect you. The regulatory power handed over to the HHS secretary seems boundless. The scores of new federal agencies, boards, commissions, panels, and programs will regulate and supervise every corner of the health care economy. Instead of the rule of law, hammered out in legislative deliberations by elected and accountable representatives, we will be governed by unelected regulators.

The rules may be applied one way to you and another way to others who are in the same situation as you are. Will you have the clout to get a waiver from the rules? Maybe you will and maybe you won't, but either way, the regulators will be in control—and they won't be accountable to you. In the crucial details of health care policy—what is or is not in your insurance policy, for example—your personal life will be directed by organizations you do not even know—people you'll probably never meet.

Some of these people work in the Center for Consumer Information and Insurance Oversight (CCIIO), created by the HHS secretary soon after the law was enacted. It has an enormous amount of power and is now housed in the Medicare bureaucracy. Yet there is no mention of this office at all in the new law. It will ensure that all health plans comply with federal marketing rules, it will enforce the insurance rate and premium rules it sets, and it will oversee state health care exchanges and administer other new programs, such as the temporary high-risk-pool program. Yet none of this enormous authority was granted directly to this agency by Congress. Instead, Congress simply handed over broad discretion to the secretary of HHS to implement ObamaCare, resulting in well over a thousand directives in the legislation that begin with the words "The Secretary shall . . ."

If you think there is something wrong with being governed like this, you and millions of others who opposed ObamaCare are right. That's why this grim forecast of what ObamaCare might look like in the future will change, because a great majority of Americans will insist on it.

This will not stand.

"We are fast approaching a tipping point where more Americans depend on the federal government than on themselves for their livelihoods—a point where we, the American people, trade in our commitment and our concern for our individual liberties in exchange for government benefits and dependencies."[59]

—Representative Paul Ryan

IMPACT ON . . .
FAMILIES AND YOUNG ADULTS

Here are my goals—reduce costs, increase quality, coverage for everybody.[1]

—Then-senator Barack Obama
January 17, 2008, in a meeting at the
San Francisco Chronicle offices

And we'll lower premiums for the typical family by $2,500 a year. [2]

—Then-senator Barack Obama
February 23, 2008, at Ohio State University
Medical Center, Columbus, Ohio

Most people initially had high hopes for health reform. There clearly are problems that must be fixed. The president had promised that if reform passed, everyone would be able to get health insurance, costs would go down, and we would be able to keep both our doctors and our coverage. And we were told that reform would even cut the deficit and make Medicare stronger.

But the law that actually passed became a Rube Goldberg contraption that can't possibly work and that fails to meet its promises—and it will make many problems worse. Officials say it will leave at least 23 million people uninsured, it is already making health insurance more expensive, and it threatens major changes to the coverage that tens of millions of Americans have today. Seniors are frightened that its cuts to future Medicare spending will jeopardize their care, and taxpayers see a flood of red ink far into the future.

ObamaCare is leaving a comet tail of broken promises as it steamrolls its way through our economy and into our lives. What happened? How could there possibly be such a big gap between promise and reality?

WHAT ABOUT HEALTH CARE COSTS?

Getting health costs under control was supposed to be the main goal of health reform. Countless times during the presidential campaign and in the months leading up to the vote on Obama-Care, the president said his plan would lower health costs.

When he said that his plan "will lower premiums for the typical family by $2,500 a year," that was a specific and tangible promise. But his plan won't reduce health costs for American families. In fact, costs already are *increasing* because of early "benefits" in ObamaCare, and they will soar once the full overhaul law takes effect.

The law won't lower taxpayer spending on health care, either. At a news conference in September 2010, Mr. Obama said "we knew that" health spending would go up as a result of more people getting health insurance. But the Associated Press said this qualification was "rarely heard" during the debate as the president and his allies repeatedly said their bill would lower both costs and government spending.[3]

It was clear to us long before the law passed that those promises could not be met with the plan he was offering. The president should have known the facts, too. Experts from the independent, nonpartisan Congressional Budget Office predicted, for example, that under the law, the cost of health insurance for American families will continue its steady rise. The analysts at CBO said that millions of Americans who buy insurance on their own will pay at least $2,100 a year *more* than if the law had not passed.[4] That insurance policy is expected to cost $15,200 in 2016 with ObamaCare but would have cost $13,100 had the law not passed. So the bottom line is that you won't save the promised $2,500 a year but will pay $2,100 a year more for insurance—which misses the target by a whopping $4,600.

In addition, CBO says that if you own or work for a small business that provides health insurance, you can expect the coverage to cost $19,200 a year for a family by 2016. And remember, buying insurance isn't optional. You must buy a policy or face federal fines and penalties.

The new taxes to pay for ObamaCare's new entitlements—tax increases of $569 billion—will cost the typical family $1,000 a year in higher premiums, costs that will hit small businesses and their employees especially hard.

"Anyone who would stand before you and say 'Well, if you pass health care reform, next year's health care premiums are going down,' I don't think is telling the truth. I think it is likely they would go up."[5]

—Senator Dick Durbin (D-Ill.)

Railing at Insurers

The person putting ObamaCare into force is Kathleen Sebelius, a former governor of Kansas whom Mr. Obama appointed as his Secretary of Health and Human Services. Secretary Sebelius threatened insurance companies for explaining that health costs are rising at least in part because of ObamaCare's early mandates: providing preventive care at no cost to patients, covering twenty-six-year-old children, raising the caps on how much the companies pay for claims, etc.

While many people value these and other benefits, they cost money and can't help but increase insurance costs.

Nonetheless, she threatened the companies, saying "We will also keep track of insurers with a record of unjustified rate increases: Those plans may be excluded from health insurance Exchanges in 2014."

This threat from Secretary Sebelius is precisely why ObamaCare is so pernicious.

Columnist Michael Barone calls this "gangster government." He wrote, "The threat to use government regulation to destroy or harm someone's business because they disagree with government officials is thuggery. Like the Obama administration's transfer of money from Chrysler bondholders to its political allies in the United Auto Workers, it is a form of gangster government." [6]

"You know we're going to control the insurance companies." [7]

—Vice President Joe Biden, March 18, 2010

HELP FOR PEOPLE WITH PREEXISTING CONDITIONS

Many families have trouble getting health insurance if they have medical problems that make it likely that their health costs will be higher than normal. It's a problem that clearly needs to be addressed.

ObamaCare created a new $5 billion program called the Pre-Existing Condition Insurance Plan to help people who have been uninsured for at least six months. In the spring of 2010, the Medicare program's chief actuary predicted that 375,000 people would sign up by the end of the year. But by early November, the Department of Health and Human Services reported that just over 8,000 people had enrolled.[8] *The Washington Post* explained the real-world experience of one citizen:

> Will Wilson, 57, of Chicago said he is "really, really, really, really discouraged." After he received an AIDS diagnosis in 2002, he discovered that his insurance at the time paid only $1,500 for medicine each year. His AIDS drugs cost $3,000 a month. He ended up in bankruptcy.[9]
>
> Wilson, a tourist trolley guide, now gets help from the federal AIDS Drug Assistance Program, but he has no coverage for other kinds of care.
>
> Wilson remembers tears streaming down his face in February 2009, the night that he watched Obama vow to Congress, "Health-care reform cannot wait, it must not wait, and it will not wait another year!"
>
> Wilson became an activist for health reform, circulating petitions, going to demonstrations. And the day after the president signed the bill into law, a *Chicago Sun-Times* column quoted him as saying, "I've had a grin on my face all day" at the prospect of the high-risk pool he could join. That was before the rates were

announced in July and Wilson discovered that the premium—nearly $600 a month—"was almost as much as my rent. It was like, no way! I was floored."

Even with $5 billion in federal subsidies, the insurance is still too expensive for many people to afford. Washington's response is the same as always: pouring more taxpayer money into subsidies to try to lower the cost of the expensive insurance. What it should have done instead is provide money to the states to either create or boost their own high-risk pools so they can tailor costs and benefits to their own citizens. This is one more example why a one-size-fits-all government plan doesn't work in health care.

LOSING THE COVERAGE YOU HAVE NOW

One analyst, Alissa Meade from McKinsey & Company, knocked the socks off insurance company executives late in 2010 when she said the new health overhaul law could mean that "something in the range of 80 to 100 million individuals are going to change coverage categories in the two years" after the insurance mandates take effect in 2014.[10]

Many will keep their employer coverage but their policies will change. Some will be forced to get their insurance through new (and untested) state health insurance exchanges. Others will go onto Medicaid—arguably the worst health program in the country. This would be an extraordinary disruption of coverage that will cause widespread outrage.

Ms. Meade predicted that 30 million to 40 million people may just decide not to buy coverage. This would be a new outlaw class of Americans—otherwise law-abiding people who would deliberately decide to break the law and not purchase the mandatory coverage, opting to pay the fine instead. The fine for individuals will be $695 a year in 2016, or 2.5 percent of income, whichever is greater. Most employers will pay fines of $2,000 per worker if they don't offer coverage—and sometimes more.[11]

This is unbelievable! Is ObamaCare going to create a nation of outlaws? You can't have a civil society and a functioning democracy when laws are so onerous that millions of people decide to break them.

But, as noted, even the government expects this to happen: the CMS actuary predicts that the government will collect about $120 billion over the next decade in fines and penalties from individuals and businesses who don't comply with the federal mandate to purchase insurance.[12]

MORE BAD NEWS

The administration has trouble keeping up with the avalanche of bad news coming as a result of ObamaCare.

The Principal Group, an insurance company operating out of Iowa, said it is dropping health insurance from its roster of products, which *The New York Times* calls "another sign of upheaval emerging among insurers as the new federal health law starts to take effect."[13] The company had covered about 840,000 people who receive their insurance through an employer. So 840,000 Midwesterners are likely to lose the coverage they have.

Principal is just one in a long list of insurers to announce plans to drop health coverage. It may have dropped it anyway, but ObamaCare likely accelerated the decision.

Lower-income workers are also threatened. Entry-level and lower-wage employees are often offered policies that are more like "starter insurance" that allow them to go to doctors, get medicines, and obtain preventive care and even some hospitalization. But these "mini-med" policies don't fit with ObamaCare's demands for expensive, comprehensive insurance. More than seven hundred companies, states, and unions got waivers to be exempt from the law so their employees didn't have to lose this coverage—at least for a year.

Liberals in Washington want nothing to do with these mini-med policies. They think that all workers must have Cadillac coverage—whether they or their employers can pay for it or not, and even if the policy costs as much as a worker's entire annual salary.

COLLEGE STUDENTS AND YOUNG ADULTS

Young adults were incredibly enthusiastic and energetic supporters of Barack Obama in the 2008 presidential election, so it's ironic—even tragic—that they could be hurt the most by his policies.

It starts with the crushing mountain of debt they will inherit to pay back trillions of dollars from new spending on the stimulus, the bailouts, and new health care programs. On top of that, trillions of dollars in red ink will flow from existing entitlement programs, especially Medicare's $30 trillion in liabilities.

But ObamaCare doubles down by targeting young people in two big ways: not only will they be forced to buy health insurance, but they will be forced to pay extra so their parents and even grandparents don't have to pay so much.

With car insurance, you can pay less if you have a good driving record. With health insurance, you would expect young and healthy people to pay less than a sixty-year-old smoker with heart disease. But the health law calls for "rating bands," which limit how much less insurers can charge young people. That means young people will pay far more than they will cost in medical claims.

And that isn't optional. College students, young people just getting their first job, and young families struggling to make ends meet and burdened with college loans will have to have health insurance approved by the federal government or pay a fine.

Most young people understand the importance of having health insurance. But they just don't know why they should have to pay up to $4,000 a year or more for a policy most of them will hardly use. A $4,000-a-year policy could add $16,000 to the cost of a college education—further adding to their student loan burdens.

Some young adults will be able to stay on their parents' policies if they are lucky enough to have a parent who still has job-based coverage—and a job. Once the mandate kicks in, some young adults may be able to buy a less expensive high-deductible policy.[14]

CHILD-ONLY POLICIES VANISHING

Secretary Sebelius also got angry when insurers started to announce that they would no longer sell new child-only insurance policies.[15]

That's because, she told the companies, they have to offer policies that cover children whenever their parents decide to buy the insurance. And insurers have to charge the same premium whether the child is a healthy five-year-old, a troubled eighteen-year-old, or a child with health problems.

No one argues that all of these children deserve coverage. But you just can't tell companies they must sell insurance to anyone who applies, even after they get sick. The companies would soon be unable to pay claims if the bills that were being submitted cost much more than the premiums people were paying. That is just not a sustainable business model.

The result is not surprising: some employers are dropping coverage for all dependents to escape the mandate. Insurers across the country, from California and Colorado to Ohio and Missouri, are dropping child-only plans, potentially impacting tens of thousands of families. And when one company in an area announces that it is going to stop offering the policies, that sets off a cascade because no company wants to be the last one in this unsustainable market.

How is this helping?

Ironically, one of the labor unions that plowed millions of dollars into helping pass ObamaCare had to acknowledge the perverse effect of the mandate to cover dependents up to age twenty-six. In the fall of 2010, the Service Employees International Union (SEIU) told dues-paying members of its big affiliate in New York that it was dropping health care coverage for children, impacting the children of more than thirty thousand lower-wage home health care attendants.

"New federal health-care reform legislation requires plans with dependent coverage to expand that coverage up to age 26," Mitra Behroozi, the benefits manager for the New York SEIU, explained in an October 22 letter to members. "Our limited resources are already stretched as far as possible, and meeting this new requirement would be financially impossible."

But as soon as the letter hit the press, the union backed off with another news release saying that dropping the coverage had nothing to do with the new federal law. Right.

How far we've come from President Obama's speech to college students at George Mason University in March 2010. To wild applause, he pledged "to all the young people here today, starting this year if you don't have insurance, all new plans will allow you to stay on your parents' plan until you are twenty-six years old."

How far we have come, indeed.

JOBS AND LOW-INCOME WORKERS

Employers today offer health insurance voluntarily. Most believe it is a competitive advantage to attract and keep good workers, and many believe it helps to keep their workforces healthy. Many small businesses want to offer insurance, but they can't afford it.

But, as we describe in the chapter "Impact on . . . You and Your Employer," starting in 2014, most businesses won't have a choice. If they have more than fifty employees, they will have to provide health care coverage or pay fines to the federal government.

ObamaCare will hit young people entering the workforce and other lower-income workers hardest, according to Diana Furchtgott-Roth, a senior fellow at the Hudson Institute, a think tank based in Washington, D.C.

"The irony is that in the name of expanding health care coverage, the administration is making it harder than ever for unskilled workers to get started in the workforce," she wrote in a commentary after the law passed.[16]

"Low-skilled workers have some of the highest unemployment rates in America," she wrote. "But, come 2014, the new health care bill will make it harder for employers to hire low-skill workers." But, she says, "low-wage and part-time jobs will start to go, not in 2014, but now."

She says that "firms will have an incentive to become more automated, or machinery-intensive—and hire fewer workers. Fast food restaurants could ship in more food and have it reheated, rather than cooking it on the premises. Department stores could have fewer sales clerks and more price-scanning stations, so that shoppers could scan labels for prices rather than asking sales assistants." She says that ObamaCare is already drying up jobs for those workers and cited an example of her local CVS drugstore: just in time for Christmas, it replaced live check-out clerks with automatic scanners.[17]

The bottom line: "With the health care bill, Washington is condemning more unskilled Americans to the ranks of the unemployed," Furchtgott-Roth wrote. Some people may find that they have neither a job nor health insurance.

"Our economy simply cannot afford this unprecedented, unconstitutional power grab by the federal government."[18]

—Rep. John A. Boehner

Under the new law, employers with more than fifty workers will either have to offer health insurance or pay an annual penalty.[19] The basic penalty for full-time employees is $2,000 per worker. For part-timers, employers will pay $2,000 for each "full-time equivalent" worker.

"Small enterprises with 50 employees or fewer will be the big winners," Furchtgott-Roth wrote. "If they don't hire too many workers—another government-induced disincentive for hiring in this weak labor market—and stay within the 50-person limit, these firms won't have to provide health insurance and will have a cost advantage over the others." But it also means that those companies won't be able to grow and hire new workers without being hit with the full force of the health care mandates.

UNINSURED

One of the biggest heartaches of liberals is that even spending $1 trillion—or $2.3 trillion if you are honest about the true ten-year cost [20]—will still leave at least 23 million people uninsured.

What about their dream of universal coverage?

The law provides new subsidies for about 32 million people to obtain insurance. Families who make less than about $30,000 a year will be thrown into Medicaid, where doctors are often paid so little and the paperwork is so onerous that few can afford to see more than a few Medicaid patients, relegating them to emergency rooms for even routine care.

Another 16 million uninsured people will receive subsidies to help them to buy government-prescribed health insurance through the new state health insurance exchanges. You won't have a lot of choices because government thinks it knows best what kind of policy we all need.

How much you receive in subsidies will depend upon your income. Most people will still have to pay something—and some people will pay a lot—to get the federally mandated insurance and keep IRS agents at bay.[21] We explain more in the chapters "Impact on . . . Taxpayers" and "Impact on . . . You and Your Constitutional Rights."

YOU WILL BUY HEALTH INSURANCE!

Families will find they are paying a high price—forced to pay a high price, actually—to buy health insurance, which may become the first or second most expensive item in their budgets.

At a presidential debate in February 2008, then-candidate Obama said, "Senator Clinton believes the only way to achieve universal health care is to force everybody to purchase it. And my belief is, the reason that people don't have it is not because they don't want it but because they can't afford it."

After the campaign, the president flipped. "I am now in favor of some sort of individual mandate as long as there's a hardship exemption," he said in a CBS News interview in July 2009. He changed his mind just after Democrats in the House of Representatives unveiled legislation that included an individual mandate.[22]

Candidate Obama was right: The cost of health insurance and health care has been and will continue to be the problem, but President Obama's health care law is only making it worse.

Taxpayers will pay part of your premium, depending on your income. Families earning up to about $88,200 a year (in 2010 dollars) will get some help in paying for health insurance if they get it through the new state exchanges.

And remember, you are required to have health insurance (except for people who can prove it is a hardship or fall into an exempt category), whether you qualify for a subsidy or not.

If you don't buy health insurance, the fine will be $695 a year or 2.5 percent of your income by 2016. We already know where this could lead. Since health insurance companies will be required to sell you a policy, even if you wait until you are sick to buy it, why pay the premium when you can just fork over the much cheaper fine? One option you have is just to wait to buy insurance until you need expensive medical care.

This is clearly a prescription for a failed insurance system.

IMPACT OF REQUIREMENTS ON PRIVACY

In order to pass out taxpayer subsidies for insurance, the government and your employers are going to need to know a lot more about you. That's one of the reasons why the Obama administration plans to hire 16,500 more IRS agents.[23]

The law says that if you make less than $88,200 (in 2010 dollars) and have to pay more than 9.5 percent of your *household* income for health insurance, you can buy subsidized coverage in the new insurance exchanges, starting in 2014.

But if you are married and get health insurance through your job, how is your employer going to know how much your husband or wife makes? And what if you get separated or divorced and your household income changes? Your employer will need to know that or possibly face fines. And state exchanges are going to need to know your income and employment status to calculate whether you are eligible for subsidies. If you make too much, you will have to pay it back.

The health overhaul law also calls for group health plans to have electronic systems.[24] The health care sector would be much more efficient if more information were available to our doctors electronically, but the issue, once again, is concern about the government potentially having access to detailed information about our health care.

INDIVIDUAL RESPONSIBILITY

Representative Paul Ryan wrote an article for his hometown newspaper, the *Milwaukee Journal Sentinel*,[25] the day President Obama signed the bill into law. "The yearlong partisan crusade—right through its ugly conclusion—revealed that this debate was never about policy but rather a paternalistic ideology at odds with our historic commitment to individual liberty, limited government and entrepreneurial dynamism," he wrote. "The proponents of this legislation reject an opportunity society and instead assume you are stuck in your station in life and the role of government is to help you cope with it. Rather than promote equal opportunities for individuals to make the most of their lives, the cradle-to-grave welfare state seeks to equalize the results of people's lives."

Health care reform, and for that matter our society and our economy, won't work if they aren't based upon individual responsibility. Yes, we are a compassionate nation, and we want to—and do—take care of people who are less fortunate than us. But we must care for ourselves and our families first, or our civil society just can't work.

In government-controlled systems, individual responsibility is replaced by political authority. This ultimately leads to rationing of health care by people who will never even know our names.

The challenges with ObamaCare will become even more intense going forward. As medical technology advances and there are more and more options for health care goods and services that are more and more expensive, the pressures will grow for someone to make decisions about access to these technologies. It can be either government bureaucrats—or us.

IMPACT ON . . .
SENIORS

Medicare is an indispensable part of American health care. The program provides health insurance for 47 million senior citizens and people with disabilities. Medicare is popular because seniors—the largest users of American health care—rely on it to get the care they need, and it's an important part of the safety net for people with disabilities.

But Medicare faces serious challenges. The program is simply not financially sustainable in its current form. It is more than $30 trillion in the red.[1] Health care costs continue to rise faster than inflation, and millions of baby boomers will be signing up for Medicare in waves as they turn sixty-five over the coming two decades. We must make changes so Medicare can be safe and solvent both for today's Medicare beneficiaries and for future retirees. That requires real and far-reaching reform that will make Medicare more efficient and less costly.

But ObamaCare's authors did not approach reform from that perspective. Instead, their goal was to create a new universal health care entitlement, and they saw Medicare as a piggybank to pay for it. The result is that the changes that ObamaCare made to Medicare do very little to improve the program's financial outlook, and the cuts to the program put at risk access to care that millions of seniors take for granted today.

For example, ObamaCare will assign seniors to new and untested plans with no guarantee that the government will get their permission to do so or even inform them that the assignment has occurred. It will create unaccountable boards that will make decisions about how much—or how little—their doctors and hospitals will be paid. And it will make cuts throughout the program that will limit their access to care. It's not surprising, therefore, that seniors are upset about ObamaCare. They should be!

THE GOVERNMENT'S FAVORITE HMO: ACCOUNTABLE CARE ORGANIZATIONS

ObamaCare's apologists believe that the government will be able to reorganize the $2.7 trillion health sector and make it more efficient without harming the quality of care. They believe this even though the federal government has never been able to control costs in the price-controlled, micro-managed health programs it has been running for nearly a half century.

Nonetheless, their faith in government-run health care is what led them to embrace an untested new idea—Accountable Care Organizations, or ACOs—making this a centerpiece of their new bureaucratic management of Medicare.

ACOs are a concept created by a number of academic researchers.[2] The idea is to create new organizations in which doctors and hospitals would work more closely together. They would voluntarily create new legal entities that ACO advocates say will be more "accountable" for providing care in doctors' offices, hospitals, and other care settings. Physicians and hospitals would share the revenue and keep some of the savings if they can provide the care at less cost than what traditional Medicare would pay.

In the 1990s, when managed care, particularly HMOs, was dominant, there was a backlash, and Congress considered passing what was then known as a patients' bill of rights to prevent HMOs from unfairly denying care to their enrollees.

Ironically, ACOs strongly resemble the managed care plans of the 1990s. They are expected to provide all of the medical care a patient needs within a budget, and if they do so while cutting costs, they can keep part of the savings as a profit. There will, of course, be a very strong incentive for the plans to find ways to provide less care to patients.

But this time, it's the government itself that is pushing managed care, not private insurers, as a way to cut costs.

Moreover, instead of providing protections to seniors, the government can leave them out of the equation entirely.

At least as conceived by its authors and as written in Obama-Care, there is no requirement at all that Medicare patients have to give consent to be enrolled in an ACO. Here is what the law says:

> (c) ASSIGNMENT OF MEDICARE FEE-FOR-SERVICE BEN-EFICIARIES TO ACOS.—The Secretary shall determine an appropriate method to assign Medicare fee-for-service beneficiaries to an ACO based on their utilization of primary care services provided under this title by an ACO professional described in subsection (h)(1)(A).[3]

In plain language, this means that the government will have the authority to "assign" Medicare beneficiaries to ACOs without necessarily even telling them or asking their permission. They can make the assignment based entirely on whether or not a patient's primary doctor has decided to work through an ACO. If the doctor signs up with an ACO, the doctor's patients will go along, whether they want to or not.

ACO advocates have designed it this way on purpose. They are worried that asking seniors to make a choice for themselves will mean that many of them would decline to participate in ACOs if asked. They won't trust them.

But forcing seniors into ACOs without their consent is a terrible way to run a program. It will create resentment and fear once seniors figure out what is going on with their health care. The only way ACOs can control costs is by steering patients to some specialists and not to others or denying access to some tests, treatments, or medicines. That kind of restricted choice, without the consent of patients, is bound to create tensions.

It's unlikely to work anyway. The ACOs don't take away seniors' right to see whatever doctor they choose. The ACOs will just give their primary doctors an incentive to refer them to some specialists but not others. There is no penalty for Medicare patients if they don't stick with the ACO. But most people follow their doctor's recommendations, and that is what the government is counting on.

ACOs are likely to become one more failed government experiment. They can't possibly deliver significant savings without altering how patients experience the Medicare program. But if Medicare patients see the ACOs as denying them care without their consent, they will rebel.

THE UNACCOUNTABLE IPAB CAP ON MEDICARE

The government's expanded control over medical decisions is just beginning. Get ready for the Independent Payment Advisory Board, or IPAB. The former director of the White House Office of Management and Budget, Peter Orszag, speculated just after ObamaCare was signed into law that the IPAB just might be viewed decades from now as the most important and far-reaching change enacted in the entire health care reform legislation. He might be right—despite the fact that it got almost no attention in the debate before the legislation was passed.

The IPAB is a fifteen-member independent panel—to be appointed by the president and confirmed by the Senate—whose main job is to enforce a limit on how much the government spends on Medicare each year.

That's right: Medicare spending is now officially capped. Even most ObamaCare advocates don't seem to know this. Perhaps it's just too hard to believe that a Democratic Congress, prodded by a Democratic president, actually voted to cap spending on a cherished entitlement. But it's the truth.[4]

The IPAB has the authority to make recommendations to cut Medicare's spending to stay within the annual caps. Those recommendations will automatically go into effect unless Congress overrides them.

But the IPAB is strictly limited in what it can recommend and implement. The only thing it can do is cut Medicare payment rates to those providing services to the beneficiaries.

At first, the IPAB mandate might seem friendly to Medicare enrollees. But that's an illusion.

The federal government has never shown any ability to cut costs in Medicare by carefully targeting waste and inefficiency. Politicians can never decide on the criteria for making such cuts.

Instead, the way Congress and Medicare's regulators have cut costs is simply to pay providers less, without regard to quality or efficient performance. Tellingly, that's exactly how ObamaCare achieves most of its Medicare cuts.

When push comes to shove, the IPAB will almost certainly fall into the same trap. To cut spending fast and with certainty, the easy solution will always be to impose across-the-board reductions on all hospitals, doctors, and others providing services or supplies to patients.

But, as Medicare's top financial expert has said repeatedly in his statements about ObamaCare, there are real limits to how low Medicare's payment rates can fall. At some point, those who are providing care or making medical products available to patients will drop out of Medicare and stop supplying their services. That's what always happens with price controls. The only way to balance supply and demand thus becomes waiting lists and queues for services. Access to care will become more difficult, and the quality of care will suffer as well.

What's worse, the IPAB can impose those kinds of damaging cuts without answering to anyone. The board members are not elected by voters. They are appointed to their positions and cannot easily be removed. They will have an enormous amount of power to control the kind of health care seniors get through Medicare, but they won't be accountable to patients at all.

THE MEDICARE SHELL GAME

ObamaCare is trying to rewrite the rules of basic math. When you cash a $1,000 check, you can spend it on a $1,000 TV or on a $1,000 refrigerator, but not both. In other words, you can spend the money only once, not twice. That's common sense, of course.

But that's apparently not how ObamaCare's authors think about money. They claim the same money can be spent twice. ObamaCare will reduce the amount that Medicare will spend over the next decade by about $500 billion, and it will raise Medicare taxes by an additional $200 billion.[5] The authors of the new law claim that this money can be used *both* to pay for a new entitlement program for others and to refill Medicare's depleted coffers to pay for benefits in the future.

But, of course, if something sounds too good to be true, it probably is. Congress can play budgetary shell games, but it can't turn $1 into $2. The same money can be spent only once, and by spending the Medicare savings and taxes on something other than Medicare, ObamaCare's authors have weakened Medicare and left its future very much in question.

That's not just our opinion; that's the view of the government's top independent experts on Medicare's financial future.

The person in the executive branch who knows and understands Medicare's numbers better than anyone is the chief actuary. He is a civil servant working for the president. Since 1995, the chief actuary has been Richard S. Foster. He and his staff analyze the viability of the Medicare trust funds for the Medicare Board of Trustees.

Mr. Foster and his staff exposed the fraudulent nature of the Medicare spending claims made by ObamaCare's apologists throughout the yearlong debate. In April 2010, they wrote a memo about ObamaCare's double counting of Medicare cuts and taxes, saying:

In practice, the improved [Medicare] financing cannot be simultaneously used to finance other Federal outlays (such as the coverage expansions under the [the new health-care law]) and to extend the trust fund, despite the appearance of this result from the respective accounting conventions.[6]

In other words, because ObamaCare directs the Medicare tax increases and supposed spending savings to another entitlement program, the federal government will have no more money than it does today to pay Medicare's bills in future years. The only way to pay Medicare's bills will be with more borrowing and more government debt. That may or may not be a realistic option in the future, given that the country's debt obligations may already be reaching crisis levels. So much for making Medicare stronger!

Mr. Foster and his team were not alone in calling out the administration and congressional Democratic leaders for this shell game. So did the independent experts at the Congressional Budget Office who work for Congress. The director of the CBO, Douglas Elmendorf, said this in December 2009:

> To describe the full amount of [Medicare] trust fund savings as both improving the government's ability to pay future Medicare benefits and financing new spending outside of Medicare would essentially double-count a large share of those savings and thus overstate the improvement in the government's fiscal position.[7]

The Obama administration and its allies in Congress could have strengthened Medicare by enacting changes that would lead to genuine savings and devoting them entirely to Medicare. But that's not what they did because their main concern wasn't Medicare but ObamaCare. Consequently, your Medicare entitlement is now in worse shape than it was before this law was enacted, and the government has taken on new entitlement obligations with almost no ability to pay for them.

DRIVING YOUR DOCTORS AND HOSPITALS
OUT OF MEDICARE

In August 2010, when the Medicare Board of Trustees released its annual report on the outlook for the program, it was an upbeat assessment, based on the shell game of double counting and tax hikes.

What was most remarkable about that report was not what the trustees themselves said. What really stood out was what Mr. Foster had to say about the report. He and his staff had prepared all of the numbers for the trustees, as they dutifully do every year. Because the double-counted cuts to Medicare are in the law, they must be included in the estimates prepared by Mr. Foster and his colleagues. But in the back of the report, Mr. Foster wrote an unprecedented statement—something like an open letter to the public—in which he essentially disavowed the findings of the previous 280 pages.[8] He warned readers not to rely on the numbers in the main report because they were completely disconnected from reality. And he directed readers to an alternative analysis that he and his staff had prepared and that they believed more accurately reflects Medicare's perilous financial condition.

It's hard to overstate the importance of what Mr. Foster did in issuing this statement. He is the point person for assessing the financial status of Medicare. He works for the president of the United States. And he declared that many of the claimed benefits for Medicare from ObamaCare—claims that the president himself has repeated over and over again—are simply not true.

Not only that, Mr. Foster made it clear that ObamaCare will actually do great damage to Medicare by driving doctors and hospitals out of the program. In many cases, Medicare's reimbursement rates will no longer cover what it costs doctors to take care of patients. The result will be that many retirees will find it much more difficult to get a doctor's appointment, and they may need to travel much farther than they do today to get care.

PAYING DOCTORS AND HOSPITALS LESS AND LESS

ObamaCare makes deep and arbitrary cuts to what Medicare will pay for hospital and other services—cuts that are used to create the fiction that ObamaCare reduces the deficit.[9]

Currently, hospitals receive an increase in their payment rates each year to reflect inflation. Under ObamaCare, those increases will be cut by an amount averaging close to half a percentage point. And these cuts will occur every year—forever.

The compounding effect of cuts of this size is truly massive and entirely implausible. As Mr. Foster and his colleagues explained in their alternative analysis, they would widen the gap between the actual cost of hospital care and what the government will pay for it. Even before the cuts, Medicare's payment rates are already only about 80 percent of what private insurers must pay to secure access to care for patients.[10]

Now the gap will be much wider. Under ObamaCare, Medicare's payments will fall below 75 percent of private insurers' rates by the end of the decade, which is below what *Medicaid* will be paying.[11] That's telling because Medicaid rates are already so low that many hospitals try to avoid admitting Medicaid-eligible patients and many private physicians can't afford to see them.

Mr. Foster's analysis shows that the results will be catastrophic for anyone on Medicare. In a few years, about 15 percent of the nation's hospitals will be losing so much money from taking care of Medicare patients that they will have to stop admitting them. That could mean the difference between a patient going to the hospital near his or her home and driving two or three hours to find one that will admit him or her. The problem will only get worse in the years ahead as payment rates are ratcheted down further and further.

Even as ObamaCare jeopardizes patients' access to hospitals, it ignores the most glaring problem in Medicare, which is that payments to physicians, which are already low, are now scheduled to be cut by nearly 30 percent in coming years.

There's a certain déjà vu quality to these cuts. In the late 1980s and 1990s, the Medicare bureaucracy tried to do something similar for physicians' fees. The idea was to shift more resources toward generalists, who were thought to be undercompensated for spending time with patients, and to control overall costs by limiting the growth of payments to the growth in the size of the U.S. economy.

But from the get-go, the experiment with physicians' fees went badly off course. Indeed, it was such a disaster that it is proof positive that ObamaCare's approach to cost control in Medicare is ill conceived and dangerous to the quality of care for seniors.

With physicians' fees, after several years of study, lengthy payment regulations were issued that had immediate and profound financial consequences for nearly every practicing physician in the United States.

Unfortunately, the consequences were exactly the opposite of what was planned. Instead of encouraging more physicians to enter primary care, the Medicare physician-fee schedule rewarded more specialization. That's because the fee schedule controlled only prices, not the number of services provided to patients. As Medicare's administrators have tried to hold down costs with payment cuts, specialists have increased their share of the pie by doing more tests and procedures—at the expense of primary care reimbursement rates. The government's plan backfired entirely, and we now have a shortage of primary care doctors.

Furthermore, overall costs have never been brought under control. With the number of services rising, the Medicare formula for paying doctors has gone completely off the rails. In 2012, fees are supposed to be cut by about 30 percent—unless Congress overrides it yet again.

But the irony of what is happening seems to have been lost on ObamaCare's advocates. They don't seem to realize that their approach to cost control—bureaucratic micromanagement of payment rates—has already been tried with doctors and has failed.

Instead of learning from this mistake, ObamaCare repeats it by imposing on the rest of the medical system the same kind of arbitrary price cutting that has failed so miserably with physicians. They are claiming that the savings from such cuts—which will almost certainly be overridden by Congress in future years to avoid disrupting care for seniors—can be relied upon to pay for an expensive new entitlement program that will never be rolled back once millions of people are signed up.

Everyone now agrees that the deep physicians' pay cuts that are scheduled to occur in 2012 and beyond would be devastating for Medicare beneficiaries. Doctors are already threatening to drop out of the program in large numbers.

But that wasn't enough motivation for ObamaCare's supporters to fix the problem. Instead, they chose to spend every dollar of the supposed Medicare savings to create a non-Medicare entitlement program. The result is that all of the money is gone, physicians are threatening to abandon the program, and the rest of Medicare is being cut in ways that will drive other providers of services to stop serving patients as well.

Medicare is valuable to seniors because it gives them access to care. Instead of strengthening that access, ObamaCare dramatically weakens it. Indeed, ObamaCare fails to fix old problems that everyone knows must be fixed, even as it creates new problems that are just as bad or worse.

IF YOU LIKE YOUR MEDICARE
ADVANTAGE PLAN . . . TOO BAD

If you are a senior you might be enrolled in a Medicare Advantage plan, which is a popular choice. Medicare Advantage is the private insurance option in Medicare. You are allowed to elect, on a voluntary basis, to get your Medicare coverage through a private plan, and, when you do so, the government pays the plan a fixed amount to provide benefits for enrollees.

Medicare Advantage plans are popular because they provide additional benefits and lower cost sharing than traditional Medicare. Today, there are more than 11 million Medicare participants enrolled in Medicare Advantage plans. Seniors like the benefits the plans provide and how they are treated.[12]

But the Obama administration and its allies in Congress don't have the same favorable view of Medicare Advantage. They say that Medicare Advantage plans cost more than traditional Medicare. But that's not really true, according to the Medicare Payment Advisory Commission, which advises Congress on Medicare's operations. MedPAC, as it is known, says the cost of providing Medicare-covered benefits through Medicare Advantage plans is no more expensive than the cost of providing them through traditional Medicare. In fact, it is less expensive to provide Medicare coverage through Medicare Advantage plans that are run as HMOs than through the traditional program.[13]

This explains why the administration and many Democrats have been so adamantly opposed to allowing real price competition in Medicare between those private plans and the government-operated traditional Medicare program. They know that this competition would allow less expensive private Medicare Advantage plans to gain leverage and attract even more enrollment.

OBAMACARE'S HOSTILITY TO PRIVATE PLANS IS NOTHING NEW

This antipathy toward giving seniors real choices in Medicare was on full display in the late 1990s. At that time, the bipartisan leadership of a Medicare Commission[14] recommended moving toward a level playing field in paying for private plans and the traditional program. But Clinton administration appointees killed the idea. It was because they were worried that millions of seniors would make the choice to sign up with private insurance.

That's something ObamaCare's defenders don't want to happen either. When seniors sign up with a Medicare Advantage plan, the government loses some of its control over their health care. And ObamaCare was written to give government *more* control over health care, not less.

Medicare Advantage HMOs can offer Medicare-covered benefits at less expense than traditional Medicare despite being placed at a huge disadvantage in today's Medicare program.[15] MedPAC found that, on average, more tightly controlled Medicare Advantage HMOs provide Medicare benefits for just 97 percent of the cost of fee-for-service reimbursements.[16] These HMOs are by far the most popular form of Medicare Advantage plans, with nearly 80 percent of Medicare's private plan enrollees choosing them.

Overall, the Congressional Budget Office estimates that ObamaCare will cut Medicare Advantage plans by a total of $136 billion over ten years.[17] These deep cuts will force the plans to raise premiums and other costs to seniors and to eliminate some benefits for things such as preventive services not covered by Medicare and vision and dental care. Some plans may have to leave markets entirely because of the cuts.

The cuts begin right away, with 2011 payment rates frozen at the 2010 levels. But that's just the beginning. Between 2012 and 2017, ObamaCare will phase in a new payment formula that will dramatically lower Medicare Advantage payments in every region of the country. The deep cuts that ObamaCare makes in payment to hospitals and other medical providers in the traditional Medicare program will also result in additional cuts in what the program pays Medicare Advantage plans.

IF YOU LIKE YOUR COVERAGE, YOU CAN KEEP YOUR COVERAGE—OR CAN YOU?

Before ObamaCare, the chief actuary for Medicare expected enrollment in Medicare Advantage plans to increase to about 14.8 million in 2017. Now, however, he expects enrollment to fall to just 7.4 million in 2017, or half of what it would have been without the cuts.[18]

The Medicare Advantage cuts are thus a clear violation of the president's oft-repeated promise that ObamaCare would not force people out of the plans they have and like today. Because of ObamaCare's cuts, some 7 million–plus seniors are going to lose the coverage they have chosen for themselves, or would have chosen in future years.

These cuts will directly reduce what Medicare Advantage provides to seniors—by an average of $3,700 annually by 2017. That's a 27 percent cut.[19] The total cut to Medicare Advantage will reach $55 billion a year by 2017.[20]

The level of Medicare Advantage reduction differs substantially by region—but no region is spared. Among counties with at least 100,000 people, the smallest cut is 15 percent (in Tuscaloosa, Alabama) and the largest is 45 percent (in Ascension, Louisiana).

The cuts are based on a formula that doesn't take into consideration many important factors. For example, in some parts of the country, Medicare costs are already very low because health care is provided more efficiently there than elsewhere. But Obama-Care will cut Medicare Advantage payments in those areas very deeply too, even though other parts of the country have much higher costs. For instance, if you live in Portland, Oregon, the average cost of taking care of a Medicare patient is only $589 per month. But in Dade County, Florida, the average cost is $1,213 per month. Yet ObamaCare will cut Portland's Medicare Advantage payments by 32 percent per beneficiary while Dade County's will fall by only 18 percent.[21] There's no good explanation for imposing cuts so unfairly.

At the state level, the average Medicare Advantage cuts are just as random. The cuts range from a low of $2,020 per person (or 21 percent) in Nevada to a high of $4,693 (or 36 percent) in Hawaii.[22]

The deep reductions in Medicare Advantage payment rates and benefits will hit low-income seniors especially hard. Many retirees who have worked for large employers or for state and local governments have access to retiree wraparound plans that cover what Medicare does not. Other retirees with sufficient income can buy Medigap coverage. But lower-income seniors do not have such options. For them, Medicare Advantage has offered better coverage and lower out-of-pocket costs than traditional Medicare—and without the expense of a second premium payment. Consequently, lower-income seniors are much more likely than higher-income beneficiaries to sign up with a Medicare Advantage plan, and the cuts will hit them especially hard. A full 70 percent of the Medicare Advantage reductions in 2017 will fall on seniors with incomes below $32,400.[23]

The deep reductions will also hit minority seniors with larger-than-average cuts. Hispanic Americans are twice as likely, and African Americans are 10 percent more likely, to be enrolled in a Medicare Advantage plan than the average Medicare beneficiary. Consequently, in 2017, ObamaCare will force almost 300,000 Hispanics and more than 800,000 African Americans out of the Medicare Advantage plan they would have preferred.[24]

SENIORS REMAIN OPPOSED TO OBAMACARE

ObamaCare's apologists both inside and outside of government have been trying to convince seniors since passage of the new law that what they have done to Medicare and the rest of American health care will be good for them.

They are sparing no expense in this political sales job. Among other things, the Department of Health and Human Services sent a flyer in the mail to Medicare enrollees in the summer of 2010 boasting about the supposed benefits of the new law.

Similarly, the department has sponsored an expensive print and television advertising campaign featuring the actor Andy Griffith, apparently hoping that delivering their misleading message in a soothing, Mayberry-esque accent might hoodwink some viewers into believing what they are saying about the law.

But it's not working. Seniors are not gullible, and they know exactly what is going on. Polls show that they remain as opposed as ever to ObamaCare—and for good reason.

The truth is that ObamaCare imposes steep cuts in Medicare to pay for new federal entitlements that we can't afford. The Medicare cuts are of the kind that will leave seniors with fewer options and no real escape from the damage they will inflict. There will be fewer hospitals and doctors willing to take care of them in the future because of the deep cuts in what Medicare will pay for their services. If you are in a Medicare Advantage plan, you could very well lose your access to it and, if not your access, you will lose $3,700 in services provided to you by that plan every year by 2017.

The other provisions of the new law that the Obama administration touts—spending more for drug coverage, in particular—are worth far less than the cuts the law imposes. The Congressional Budget Office, for instance, estimates that the expanded benefits in Medicare under ObamaCare are worth less than 10 percent of the cuts it makes.[25]

In short, ObamaCare is going to reduce your choices, impair your access to care, and increase your costs—all in direct violation of what the president promised.

Many supporters of ObamaCare seem genuinely surprised and alarmed that they can't seem to convince seniors that Obama-Care is good for them. They shouldn't be surprised. It's plain that ObamaCare will worsen the health care experience for many millions of seniors. Seniors aren't going to take that lying down, nor should they. The rational response is to hold those who passed this law accountable for what they have done. And seniors started to do that with the votes they cast at the ballot box in November 2010.

IMPACT ON...
VULNERABLE AMERICANS

One of the most tragic results of ObamaCare is that it is going to make it harder for many vulnerable Americans to get health care—whether they are people with low incomes already on Medicaid, seniors in popular Medicare Advantage plans, children with health problems who are losing insurance, or Americans with health problems and diseases who are hoping for new treatments and even cures. Though many people believed that ObamaCare would improve their access to health care, they will find that the law's profound changes will lead to a cascade of unintended consequences that will make it more difficult for them to get the care they need.

It is impossible to fit the health care needs of more than 300 million individual Americans into one law with thousands of pages of legislation and regulations. But that's what ObamaCare tries to do. And many people are going to be left behind. It didn't need to be this way.

THE MEDICAID GHETTO

ObamaCare puts a huge swath of our health sector under government control. As we have seen time after time in other countries that have government-run health care systems, this leads to long waiting lines, trouble getting doctors' appointments, and limits on access to new medical technologies. All of this is already happening today in the United States in the biggest of our own government-run health programs: Medicaid.

More than 58 million people are on Medicaid today,[1] and they receive medical care that is almost entirely subsidized by taxpayers—partly with money from federal taxpayers and partly with state tax money.

For many, Medicaid is a paper promise. The law says patients can have an almost unlimited list of medical services. But in reality, many find it extremely hard to find a private doctor who can afford to see them. That's because in many states, Medicaid pays doctors much less than it costs them to see patients. Few doctors can afford to see more than a few Medicaid patients because the program pays them so little. This is especially true with specialists. One Florida doctor reported that after a long battle with the state after treating a Medicaid patient with a complex lung disease, he received a check for one cent.

About half of those who are supposed to get new health coverage under ObamaCare will get it through Medicaid, but there just aren't enough doctors to see them. Well over half of primary care doctors say they are not accepting any new Medicaid patients.[2]

Patients confined to this "Medicaid ghetto" often go to hospital emergency rooms to get even routine care. If they wait there long enough, they will eventually be seen. But at what price—both to them and to our health care system?

People on Medicaid (and the Children's Health Insurance Program [CHIP], which often pays doctors at Medicaid rates) use emergency rooms far more often than people with private insurance. And they are more likely to visit emergency rooms many times in a year.[3] Many go to ERs because they don't have anywhere else to go for care.

You would have thought that Congress would try to fix this with its major health care overhaul. But that is not what Obama-Care did. Instead, it doubled down on the problem by requiring states to add 16 million *more* people to Medicaid.

Until now, states have basically decided how much people can make and still be on Medicaid. (There are different categories that people must fit into to be eligible.) In New Jersey, for example, working parents could earn up to $36,612 in 2009 and be on Medicaid, but in Texas, the cutoff was $4,824.[4]

Under ObamaCare, individuals who make $14,404 a year and families with incomes up to $29,327 (in 2010 dollars) will get their health care through Medicaid.[5] Every state that participates in Medicaid—and all states do—will have to comply. Some states, such as Texas, are considering dropping out of Medicaid altogether because they just can't afford the added cost.

Federal taxpayers will pick up most of the tab for the 16 million new Medicaid recipients, at least at first. But many governors fear that adding these and millions more people to their Medicaid rolls will cripple their state budgets. Medicaid is already consuming a huge share of their revenues. Medicaid will inevitably take money away from spending on education, transportation, public safety, and other programs.

Fraught with Fraud

Why should we be dramatically expanding a program that is so plagued by fraud? *The New York Times* ran a series of articles in

2005 exposing the astonishing corruption in New York State's Medicaid program:

- A Brooklyn dentist—who has since been indicted—billed for 991 fillings, cleanings, and other dental procedures supposedly done in one day in 2003, costing Medicaid a total of $63,967. That dentist's payments from Medicaid that year totaled $5.4 million.
- The state paid $316 million for private ambulance transportation for Medicaid patients. It paid $36 for a taxi trip that would cost $2 on a bus—in a city that has the country's best public transportation system.
- Of the 400 million claims that New York Medicaid paid in 2004, state officials found just 37 cases of suspected fraud. Yet experts estimate that at least 10 percent and likely much more of the $45 billion the state spent on Medicaid in 2005 was stolen or wasted.[6,7]

It was shocking but not surprising, then, to learn that the average cost per Medicaid patient in the state is $10,600 a year. That's enough to buy a good private health insurance policy, even in New York's incredibly expensive market.

The series got the attention of former governor George Pataki, who quickly appointed an independent inspector general to track fraud and abuse in the state Medicaid program. In the three years leading up to 2009, the unit recovered nearly $660 million and landed hundreds of convictions and millions in restitution in New York.[8] But it's still just the tip of the iceberg. Medicaid has too long been at the bottom of the priority list for policy makers and investigators. It needs to move to the top.

Making It Harder to Get Care

The fraud is awful and a tragic waste of taxpayer dollars. But the problems don't end there. The huge expansion of Medicaid will make it even harder for the patients who are already on Medicaid to get the health care they need.

Even before ObamaCare was enacted, the dean and CEO of Johns Hopkins Medicine, Edward Miller, warned that putting millions more people on Medicaid would mean overwhelming demand for medical centers like his that treat a large number of low-income patients.

Dr. Miller wrote a commentary article in *The Wall Street Journal* in December 2009 entitled "Health Reform Could Harm Medicaid Patients."[9] He warned that this large Medicaid expansion could have "catastrophic effects on those of us who provide society's health-care safety-net."

Hopkins serves tens of thousands of poor, disadvantaged people, including 150,000 people in Maryland's Medicaid program. Hopkins has worked very hard to create programs to provide quality care, ranging from routine care at clinics to sophisticated hospital treatment for patients with serious and complex medical problems.

"The key fact is that for years the state did not cover all the costs [of] our Medicaid program," Dr. Miller wrote. Johns Hopkins lost more than $57.2 million treating Medicaid patients between 1997 and 2005. The state had added thousands more people to Medicaid "whose costs were not completely covered by the state." Then Maryland expanded Medicaid *again* to cover more people, and Johns Hopkins lost another $15 million in just the first nine months. There is just no way the system can handle the huge wave of new patients that is coming with ObamaCare's Medicaid expansion.

Given time, Dr. Miller says, Hopkins could work with other medical facilities to create a system of care for thousands more patients. But if 16 million more people are added to the Medicaid rolls nationwide in 2014, it could completely overwhelm the safety-net system among his and other hospitals and clinics around the country. And given Medicaid's abysmally low payment rates, it is unlikely that private doctors will be able to afford to take much more of the exploding caseload.

What this means is that patients who are already on Medicaid will be competing for medical services with millions more patients who are being added to the program. And the most vulnerable patients who have the greatest needs are likely to have the hardest time getting care.

There is a better way: allow people on Medicaid the option of private insurance. Instead of a paper promise, lower-income people could get private coverage through competing plans, which could better manage care and allow people to be seen in doctors' offices rather than in expensive hospital emergency rooms. We write more about how to do this in our chapter "What We Should Do Instead to Get Reform Right."

Canary in the Coal Mine

Massachusetts is a canary in the coal mine for health reform and offers a lesson for other states. Before the state enacted its own version of reform in 2006, it had the highest health costs in the country. But health costs continue to soar in the Bay State, and Medicaid spending is already choking its budget.[10] Today, three-fourths of the people who have become insured as a result of the Massachusetts law[11] are getting taxpayer-subsidized coverage, through either Medicaid or its version of an exchange.[12]

A recent report on the status of Massachusetts' health care reform efforts said, "Medicaid is gobbling up more and more of the state budget, a trend that has been going on for many years. . . . [It's] devouring new state revenues and leaving other services in areas like public safety, human services, education and local aid, subject to continuing budget cuts."[13]

A separate report shows that hundreds of millions of dollars that were supposed to go to improving public schools in Massachusetts have instead gone to pay for expensive health insurance for teachers.[14] The rise in health insurance premiums has "completely consumed the increased appropriations for education and then some," according to the report from the Boston Foundation. "These cost increases are huge, and they're affecting kids."

Health care spending, especially spending driven by entitlements, can turn into a voracious monster. At the federal level, with the enormous burdens of Medicare and Medicaid, defense spending will certainly face the same fate that education spending is facing in Massachusetts.

By the end of this decade, experts estimate, 84 million people in total will be on Medicaid—a program that was originally designed for the poorest Americans.[15]

ObamaCare severely stresses what should be a safety-net program. This is not reform. Why would we put so many people into a program where people already have such a hard time getting care?

MEDICARE

More seniors are reporting that they have trouble finding a physician who will take new Medicare patients. To provide options, Congress created a new program in 2003 called Medicare Advantage, which lets seniors pick a private health plan to provide their medical care. Many Medicare Advantage plans resemble the insurance many people had with their jobs and offer more benefits and better coverage than traditional Medicare does. There are even special Medicare Advantage plans for people with specific illnesses, such as diabetes.

Private plans in Medicare are so popular that nearly one-fourth of all Medicare beneficiaries have voluntarily signed up for them.

So what did ObamaCare do? It slashed funding for that program.

The president and the Democratic leaders in Congress who passed this law apparently can't abide having any private companies involved in providing health care. They took $145 billion out of Medicare Advantage to create new entitlement programs for working Americans.[16] As noted, the chief actuary for Medicare says more than 7 million people will lose the coverage they have now with Medicare Advantage as a result of the payment cuts, as we explain in our chapter on "The Impact on . . . Seniors."

One of the first announcements came from Harvard Pilgrim Health Care in September 2010. It announced that it was getting out of the market for Medicare Advantage in response to the cuts coming to the program. That meant that 22,000 New Englanders would have to find other coverage.[17] Many of them are lower-income seniors who don't have retiree coverage from their jobs or who can't afford expensive Medigap insurance. So ObamaCare is harming senior citizens presently on Medicare Advantage, who will now be forced back into traditional Medicare, where they will find it harder and harder to find a physician to see them and who will likely face higher costs as well.

How is this progress?

Those who wrote the health care law think they have a solution for this in the form of "Accountable Care Organizations." This is a new entity that does not yet exist, yet many people are hoping that ACOs will be the money savers in ObamaCare. We explain more about them in the chapter "Impact on . . . Seniors."

ACOs are a reinvention of managed care: doctors, hospitals, and other health care providers will form new organizations to provide care to seniors on Medicare. Done right, ACOs could be a way to get coordinated care, avoid duplication and unnecessary services, and make care more efficient and less costly.

Or they could be an invisible way for these new oligopolies to make money by hiding from seniors the fact that doctors will have an incentive to send them to less expensive specialists—or no specialist at all—or to skimp on care in other ways. We don't yet know. But Dr. Scott Gottlieb, a physician who is also a scholar at the American Enterprise Institute, says:

> ACOs are no miracle cure. If they squeeze out insurers, that will invariably limit the choice that patients have on the new exchanges. People will be tied to their local community for most if not all their health care. Going "out of network" will be hard if not impossible. . . .
>
> Since the ACOs will have local monopolies, they'll also have little incentive to compete for more patients in an open marketplace. Yet this is the only incentive that would spur an ACO to truly innovate and improve its delivery of medical care and offer better services. . . .
>
> If the ACOs squeeze out this competition, the result will be a de facto "single payer." Every market will be controlled by a single ACO, which will in turn answer directly to the feds, with government regulators setting the prices and all the terms of care.[18]

Not the outcome we would have wanted.

THE MOST VULNERABLE: THE UNBORN

The final hurdle that stood in the way of passing the law in March 2010 involved the very real fear that the legislation would allow taxpayer funding for abortion. A group of House Democrats held out until the final hours of the debate, insisting they wouldn't vote for the bill unless the legislation was changed.

A strong majority of Americans opposes using federal taxpayer dollars to finance abortion. Yet the legislation could allow the newly created government insurance plans to cover abortion and allow companies that receive federal funds to offer policies that include abortion coverage.

But Nancy Pelosi wouldn't change a word in the bill at that point, knowing that would mean it would have to go back to the Senate where there no longer were 60 votes to pass it. Hours before the final vote, the president persuaded then-representative Bart Stupak, a prolife Democrat from Michigan, and several of his colleagues to accept a presidential Executive Order banning the use of taxpayer dollars for abortion. No one believed it was worth more than the paper it was written on, but Stupak and his colleagues jumped on board, saying they would vote yes on ObamaCare.[19]

An Executive Order is not a substitute for a statutory provision, cannot remedy the law's statutory deficiencies in separating taxpayer funding from abortion, and is unlikely to withstand a legal challenge at any point where the order and the statute appear to be in conflict. Only a permanent, government-wide ban, modeled on the Hyde language, would guarantee the separation of taxpayer funding and abortion. The Hyde Amendment, named after former congressman Henry Hyde (R-Ill.), is a legislative provision that bans spending federal funds to pay for abortions primarily in the Medicaid program. It has been added as an amendment to appropriations bills each year since 1976.

ARE THERE DEATH PANELS IN OBAMACARE?

In the summer of 2009, town hall meetings were erupting all over the country as people, especially senior citizens, feared that the bill would create "death panels."

Are there death panels in ObamaCare? The answer is no. But a small provision in an early version of the legislation set off fears about government's role in decisions involving end-of-life care.

The furor began over a section in the House bill written by Representative Earl Blumenauer (D-Oreg.) to allow Medicare to pay doctors for counseling sessions to address end-of-life issues. Doctors would answer questions about "living wills" where people write down in advance what they want done if they are too ill to make their own decisions. The doctor would also talk about making a close relative or a trusted friend a health care proxy, provide information about medications for serious pain, and talk about hospice as an option for people with terminal illnesses.

These are all issues that people should discuss with their doctors and families so their wishes are respected. But the fear was that government will encourage people to sign a document that would as, President Obama put it, "Pull the plug on grandma." Here's how the president explained the death panel furor in a town hall meeting in Portsmouth, New Hampshire, on August 12, 2009:

> THE PRESIDENT: Let me just be specific about some things that I've been hearing lately that we just need to dispose of here.

The rumor that's been circulating a lot lately is this idea that somehow the House of Representatives voted for "death panels" that will basically pull the plug on grandma because we've decided that we don't—it's too expensive to let her live anymore. (Laughter.) And there are various—there are some variations on this theme.

It turns out that I guess this arose out of a provision in one of the House bills that allowed Medicare to reimburse people for consultations about end-of-life care, setting up living wills, the availability of hospice, et cetera. So the intention of the members of Congress was to give people more information so that they could handle issues of end-of-life care when they're ready, on their own terms. It wasn't forcing anybody to do anything. This is I guess where the rumor came from.[20]

Because of the public outcry, the controversial provision was pulled from the final bill. But Representative Blumenauer didn't give up and kept looking for new ways to achieve his goal.

By the end of 2010, he had found it. The Obama administration decided to issue a regulation to allow end-of-life counseling in Medicare as part of the "wellness" provisions in the law.

The new regulation allows Medicare to cover "voluntary advance care planning." Doctors will be paid to talk with patients about preparing an "advance directive" to give guidance on how aggressively they want to be treated if they are so sick they can't make decisions or communicate their wishes.

"It will give people more control over the care they receive," Mr. Blumenauer said in an interview with *The New York Times*, which reported the story. "It means that doctors and patients can have these conversations in the normal course of business, as part of our health care routine, not as something put off until we are forced to do it."[21]

He's absolutely right. But what made people upset this time was that the Obama administration was trying to make an end run around the legislation and push the end-of-life counseling through by regulation. People were angry about the lack of transparency. Representative Blumenauer reinforced their suspicions when he sent an e-mail to supporters to alert them, but he advised them to keep quiet lest the cries of death panels erupt again.

> "While we are very happy with the result, we won't be shouting it from the rooftops because we aren't out of the woods yet," Mr. Blumenauer's office said in an e-mail in early November to people working with him on the issue. "This regulation could be modified or reversed, especially if Republican leaders try to use this small provision to perpetuate the 'death panel' myth."
>
> Moreover, the e-mail said: "We would ask that you not broadcast this accomplishment out to any of your lists, even if they are 'supporters'—e-mails can too easily be forwarded."
>
> The e-mail continued: "Thus far, it seems that no press or blogs have discovered it, but we will be keeping a close watch and may be calling on you if we need a rapid, targeted response. The longer this goes unnoticed, the better our chances of keeping it."[22]

But *The New York Times* did find out about it and ran it as the lead story in the paper the Sunday after Christmas, creating a new uproar. The next week, the Obama administration announced that it was withdrawing that portion of the new Medicare rule.[23]

This raises the question of what other rules will go into effect without receiving the bright light of public and media scrutiny.

These are profoundly serious and sensitive issues. Decisions about end-of-life care are emotional and complex, and it's extremely beneficial to think about what we want done for ourselves and our loved ones before a crisis. What frightens many Americans is getting government involved. There is clearly the potential for a conflict of interest if the federal government is paying medical professionals for end-of-life counseling and also paying for our end-of-life health care. And there will be other issues: many Americans fear that, as government tries harder and harder to control health spending, unelected officials could recommend or even implement policies not to pay for a particular medical device or a drug they think is too expensive.

These fears are not fanciful. We can look to Great Britain to see where this leads. Some doctors simply don't tell patients in the government's single-payer health care system about advanced treatments that are available if the government doesn't pay for them. It becomes rationing by silence.

Government and Decisions About Medical Benefits

There are many ways that the overhaul law gives government officials power over medical decisions.

President Obama repeatedly has said that his health care reform plan "will not tell you which doctors to see or what treatments to get. . . . No government bureaucrat will second-guess decisions about your care."[24]

But ObamaCare does indeed create the infrastructure for that kind of intervention. Many of the 159 new boards, commissions, and programs would move key decisions away from doctors and patients and toward government agencies and bureaucracies. They could regulate and micromanage virtually every aspect of the U.S. health care sector. And they will have new and untested powers to make decisions about medical treatments impacting all of us.

This would mean government could indeed be "second-guessing" your medical treatments. The president did not help his own case when he talked about "a blue pill and a red pill" during a White House forum and suggested that the blue pill is the better choice if it is "half as expensive as the red pill."[25]

A European Model

One of the ideas Congress borrowed from European health care systems is "comparative effectiveness research" to gather information on the scientific merits of competing medical treatments. Basing medical decisions upon proven evidence is, of course, a good thing. Doctors are trained to use the best scientific evidence in making medical decisions for their patients. But many people are rightly concerned about centralized government control over research and having government officials or their agents choosing one medical treatment over another.

One result has been to slow medical research, centralize government control over what treatments are available to patients, and often deny access to new treatments.

Ignoring this experience, Congress barreled forward. In the first days of the Obama administration, Congress approved $1.1 billion in the stimulus bill to create a new government agency called the Federal Coordinating Council for Comparative Effectiveness Research.[26]

Fifteen federal government officials were appointed in March 2009 to launch the program. These officials held a series of "listening sessions" to gather testimony on government priorities for future investments in medical research.

Comparative effectiveness reviews could focus on making information about medical treatments available and then allowing doctors and patients to make decisions. That is what *should* happen. But that is not what experience shows *does* happen. It's often used to restrict access to medical care that the government believes is too expensive.

A different government agency, the U.S. Food and Drug Administration, did just that in December 2010 when it pulled back its approval for Avastin—a drug for advanced breast cancer—with at least the appearance that the government believed it was too expensive.[27]

Many people are rightly worried that ObamaCare will try to replace the experience, wisdom, and knowledge of physicians with bureaucratic decisions that will reduce medical treatments to cookie-cutter formulas. At particular risk are patients with rare diseases who fit outside the norms for standardized care.

More than a quarter of all Americans suffer from chronic conditions, such as arthritis, asthma, diabetes, or heart disease. Care for these patients consumes about two-thirds of the nation's health care budget. Giving government a bigger role in paying for medical services and health insurance—as ObamaCare does—also gives it a bigger say in what treatments we will or will not get. And if government is trying to cut down on health care spending, it's the most vulnerable Americans who are at the greatest risk.

"I think we have to worry very significantly about innovation."[28]
—Delos (Toby) Cosgrove, CEO, Cleveland Clinic

What Is Your Life Worth?

The United Kingdom has a true government-run health care system. It has a record of denying access to the newest drugs, often deciding they just aren't worth the cost. The board that does the rationing is called the National Institute for Health and Clinical Excellence (NICE).

In order to determine whether a new medicine or other therapy will be available to patients through the government's single-payer health care system, bureaucrats put a price on the value of a year of life of citizens.

The concept is called a "quality-adjusted life year," QALY for short, and get ready to hear a lot about this. This is part of the formula the British use to make sure tax dollars are being spent on the most cost-effective treatments.

NICE weighs medical results against costs to determine the "comparative value" of different medical treatments. The British government has determined that a year of life is worth about £20,000 to £30,000—to it.[29] (This is about $32,000–$48,000 U.S. dollars.) Though it's simple to say how much a treatment costs, it's much more difficult to place a monetary value on the months or years it can add to a patient's life.

Reducing human beings and their pain and suffering to dollars and cents is a difficult job, but proponents argue that faced with limited resources, health plans need some way of making decisions about what treatments are worth paying for.

But try telling that to the families of thousands of Alzheimer's patients in the United Kingdom who lost access to a promising medicine when NICE decided it was too expensive. Or the kidney cancer patients who lost access to groundbreaking new drugs when NICE decided the benefits (six more months of life on average but years more for some) did not exceed the costs (roughly $50,000 per patient per year).

The same thing happens throughout Europe. A study from the Institut économique Molinari in Belgium says that approval processes in Europe are increasingly "tough, heavy-handed and costly. . . . Despite the best intentions, the inevitable consequence of these regulations is to push up the cost of innovation substantially, to undervalue its benefits and to reduce the number of new products by making certain projects unprofitable."[30]

Healthy people may never hear about NICE or QALY. But for people with fragile health, these government formulas pose a serious risk.

Section 6301 of the final Obama health care law governs the use of comparative effectiveness research. It specifically forbids the Patient Centered Outcomes Research Institute to develop or use a "dollars-per-quality adjusted life year." It also expressly forbids the secretary to base coverage or reimbursement decisions solely on the basis of this research or to make value judgments on the basis of age. Congress will be watching closely.

Hopes of Lowering Spending

The hope of the bureaucrats in charge of implementing Obama-Care is that its new government agencies can help to lower health care spending.

But Professor Michael Schlander, a well-respected German physician, medical researcher, and economist, says that's not what happens. He found that decisions by NICE actually led to *higher* spending.[31] Experience in the United Kingdom shows that these restrictions don't save the government money. The best treatment may be more expensive at first, but it keeps people out of hospitals and from developing complications that may be even more expensive to treat. Government officials operating in silos with their proverbial green eye shades seldom see these longer-term benefits.

But rather than learning from the failures of the British health care system, the United States is marching forward to imitate the central planning so common in countries with government-run health care systems.

How important is comparative effectiveness research? Representative Nancy Pelosi's chief health policy advisor puts it first on the list of cost-control mechanisms in ObamaCare.[32] ObamaCare's authors often have said they support government using such tools to ration health care resources. A House Appropriations Committee report on the 2009 stimulus package said, "Those items, procedures, and interventions that are most effective to prevent, control, and treat health conditions will be utilized, while those that are found to be less effective and in some cases, more expensive, will no longer be prescribed."[33] Further, former Senate Democratic leader Tom Daschle, in his book *Critical: What We Can Do About the Health-Care Crisis*, wrote, "We won't be able to make a significant dent in health-care spending without getting into the nitty-gritty of which treatments are the most clinically valuable and cost-effective. . . . The federal government could exert tremendous leverage with its decisions."[34]

Policies That Don't Work for Patients

The worst thing is when the policies harm patients. Survival rates from cancer are much lower in Britain than in the United States. *The Lancet Oncology*, the prestigious medical journal,[35] reports that 84 percent of women live at least five years after being diagnosed with breast cancer in the United States. But only 70 percent of British women survive that long.

In the U.S., 92 percent of men with prostate cancer survive for at least five years, versus only 51 percent in Britain.

When British officials approved an innovative new treatment for elderly blindness as being safe, the government at first decided to pay for the treatment only after patients had gone blind in one eye. They said that a person blind in one eye can still see and is therefore not that badly off.[36]

All this has sparked a tremendous public outcry—and fortunately, British health care officials appear to have finally listened. In 2010, the government decided that NICE's policy recommendations about drugs and medical technologies now will be purely "advisory."

WHAT THIS ALL MEANS FOR THE FUTURE

Of course, we all want to know if one medical treatment is better than another. Doctors and patients need the best information to make the right choices in medical care. These are often complicated decisions that balance what treatments are available, what other conditions you might have, and even your genetic makeup. An agency that collects studies and makes the information available for research could perform an important service.

But government should be neutral as to how doctors and patients use that information. The risk is that government will use the information to reward doctors who follow its recommendations and punish those who don't.

The impact could be broader. In the name of protecting their bottom lines, health plans would have an incentive—and an "official" justification—to refuse to cover treatments and procedures that don't have an official stamp of approval.

Doctors and hospitals, fearing lawsuits, would also be much less likely to try a treatment not approved by the government institute—even if evidence suggests it might work for a particular ailment or set of patients. Bureaucracy will replace innovation.

Doctors will eventually be told they must follow the government treatment standards or risk not being paid—or even being sued—if they get creative. Our health care system will become more and more rigid as doctors fear going against the rules.

And pharmaceutical companies will be less likely to pursue research on new and potentially lifesaving drugs if they fear they will not get a positive grade. Ultimately, funds for new research will dry up.

Turning our health care sector over to bureaucrats goes completely in the opposite direction from the way science is advancing. Today your doctor can know, for example, whether or not a specific chemotherapy is going to work in treating your particular cancer—before you even get the first dose. Dr. Denis Cortese, a former president and CEO of the Mayo Clinic, has been a leader in promoting the next generation of medical care, called "personalized medicine."

We are entering what could be a transformative century in protecting, preserving, and enhancing our health through targeted treatments and preventive care.

But ObamaCare fights against this advance in medicine in two ways: First, it imposes new taxes and regulatory barriers on innovative companies that produce new treatments and cures. And second, the law gives Washington the power to set the terms and conditions of health care delivery and to decide what medical treatments and procedures will be paid for.

When government controls so much of health care spending, it can quash investments in innovation and interfere with the natural processes of scientific investigation by denying payment and therefore blocking access to the huge markets it controls.

Many investigations are better than one centralized government body in determining whether a product is efficacious. Governments too often make decisions in silos. Integrated private plans are more likely to see the overall benefit of paying for a costly drug to avert an even more expensive hospitalization.

We need a more diverse, dynamic, information-based approval system to pave the way for personalized medicine. Some patients' lives may be extended for years because they respond particularly well to a drug. Others won't. The challenge of twenty-first-century medicine is to find out why.

WASHINGTON THINKS IT KNOWS BEST

One of the most troublesome of ObamaCare's new government bodies is called the Independent Payment Advisory Board (IPAB), which, beginning in 2014, will be required to recommend ways to cut Medicare spending if it exceeds a set target. Payments will be cut for physicians, drugs, medical supplies, and eventually, hospital care.

The IPAB's recommendations will become law unless Congress passes its own proposal that achieves the same savings.

This means that fifteen people appointed by the president will be able to make cuts in payments to doctors and others treating Medicare patients, without any responsibility to answer to the people for their decisions. Congress will be able to overrule its decisions only with great difficulty. And pressures are growing within the Obama administration to grant the IPAB even more power.

At the very time that the United Kingdom is recognizing the shortcomings of government-controlled medicine and giving doctors more power over medical decisions, the United States is lurching in the opposite direction. The message of the 2010 elections was to stop this government intrusion into our freedom.

Congress is keeping a close eye on the people who are writing the mountain of regulations to put ObamaCare in place. But

Congress's powers are limited since the law is already in place and the people who wrote it are in control of the bureaucracy, writing thousands of pages of regulations to determine how the law will be implemented.

Regulators Running Amok

The New York Times published a news article in December 2010 about the mad scramble by Washington bureaucrats to write the mountain of new regulations to set up ObamaCare. "The laws were so broad and complex," the *Times* wrote, that this new army of bureaucrats has "wide leeway in determining what the rules should say and how they should be carried out."[37]

In just the first six months after the law was passed, the administration released 4,103 pages of regulations. The administration is issuing rules so fast that it isn't even giving the American people time to comment on them, no matter how much they may object to them.[38] This does not match the president's promise to provide "an unmatched level of transparency, participation and accountability across the entire Administration."[39]

And Look Who's in Charge

One of the main government agencies in charge of implementing ObamaCare is the Centers for Medicare & Medicaid Services.

The administrator is Dr. Donald Berwick. This is the man President Obama appointed to a critical post overseeing the health law's implementation. The president went around the normal process of Senate hearings and confirmation[40] and simply appointed him to the position while Congress was in recess.

It's likely the president made this end run because he and members of his administration didn't want Dr. Berwick to have to answer questions about some of the statements he has made in the past. For example, Dr. Berwick has said he is "romantic" about the British single-payer health care system. "I love it. All I need to do to rediscover the romance is to look at health care in my own country," he said at the sixtieth anniversary party for the British National Health Service.[41]

He explicitly says that he believes that the United States must ration health care and that we should do it with our "eyes wide open."[42] President Obama has renominated Dr. Berwick to head CMS so senators may soon have a chance to explore his views.

How the Law Steals Money for Research and What We Should Do to Reward Innovation in Tomorrow's Treatments and Cures

Until now, American medicine has been the finest in the world. People from all over the world come here for treatment at our state-of-the-art medical facilities. Until now, we have rewarded innovation to develop new medicines, new tests, and new medical technologies of all kinds.

Not only will the government decide what medical treatments meet its tests, but medical companies will be forced to pay higher taxes to finance ObamaCare, further draining their research budgets. We may never know about the cure for cancer or the treatment for Parkinson's disease that wasn't found because a company was forced to close the lab of a brilliant researcher.

This will affect your access to medical care today and your access to innovative cures and treatments tomorrow. This is especially true for people with serious and rare diseases. It costs, on average, more than $1 billion and twelve years to get a new drug to the market. If companies fear they will face a roadblock of regulators who could decide that their drug isn't needed, what

CEO is going to take that risk? But without this risk taking, the flow of new drugs will slow to a trickle. And people who are desperately hoping for drugs that aren't discovered yet to help them find a cure or even a treatment for their disease will be forced to wait. And wait. And wait . . .

WILL THE GOVERNMENT'S NEW RULES HARM WORKERS WITH HEALTH PROBLEMS?

Another group that should be worried about the impact of Obama-Care is people who have insurance through their jobs, especially those who have higher-than-average health costs. Amy Monahan and Daniel Schwarcz of the University of Minnesota School of Law explain why.

They say that the federal health overhaul law could encourage companies to redesign their health plans to "encourage" employees with greater medical needs to leave their employer's plan and get insurance on their own. "Although largely overlooked in public policy debates," they write, "this prospect of employer dumping of high-risk employees raises serious concerns about the sustainability of health care reform more generally." [43]

If this happens, they say, it will threaten the market for individual health insurance and state exchanges because they will have more sick and high-risk people who cost much more. As premium costs go up and up, more and more people will drop out of the plans, and eventually they will collapse.

"The real risk is NOT that employers will completely drop coverage," Schwarcz explains. "Instead, it is that employers will offer all employees revised plans that are specifically designed to induce ONLY THE LEAST HEALTHY employees to opt for coverage on the exchange." [44]

That means if you have a job with good insurance now, you could be shoved out of that coverage and into the state exchanges, where government bureaucracies will have much more control over your access to health care.

PRIVATE-SECTOR INNOVATION

Innovations in the delivery of medical care and options for health insurance could also be crippled by the huge regulatory reach of the health care overhaul law.

This will further discourage employers that have been leaders in developing health insurance policies that engage their employees as partners in managing their health care and health care spending.

Over the last decade, competitive pressures have forced private companies to develop new solutions to health care needs while driving down costs and expanding access to quality care in the process. Companies have developed thousands of new health insurance models. For example, 10 million people have Health Savings Accounts, which give them a chance to save money, tax-free, for routine medical expenses while making sure they are covered for large medical costs as well as preventive care.[45] Walmart allows employees to customize their health plans to fit their budget and their needs. The cumulative result has been a leveling of employer health costs over the last five years, keeping employers in the game and giving them more options.[46]

Competition also has forced prescription drug prices down. In 2006, Walmart began offering thirty-day supplies of several hundred generic drugs for just $4. Competitors quickly followed suit. Today, many major retailers and pharmacies offer even better deals. Instead of suffocating this progress with a mountain of new rules and regulations, the government should get out of the way and encourage more of these private-sector initiatives.

Under a government-controlled system where bureaucrats dictate every detail of benefits and payments, the incentives for such innovations would vanish. Instead of innovating, companies will be trying to jump through the thousands of new regulatory hoops to satisfy regulators rather than customers.

THE PATH FORWARD

Health care in the United States is far from perfect. But before upending it, Congress should take a long look at how such private initiatives are solving some key health care problems. It's imperative that health care reform enhance the private sector's ability to serve patients—not get in the way.

The medical profession has been speeding toward patient-centered medicine with microtargeting of treatments tailored to patients' individual genetic codes. Advances in medical science must be allowed to continue without being blocked by regulatory obstacles and restrictive payment systems. Continued innovation is vital to progress in health care, but it will be crushed by Obama-Care's rules, regulations, and restrictions.

Treating patients with rare diseases and multiple problems is enormously complex. Doctors need to take into account our individual differences and how we might respond to different drugs and treatments. Standardized care is incompatible with personalized care. Decisions based upon "averages" harm patients who do not respond well to standard care. There is simply no way government bureaucrats can understand our particular needs. Too often in government-controlled health systems, the unique medical needs of individual patients are ignored. And politicians and government officials are inevitably subject to political influences. Patient groups and companies with the most political power— and the money to hire the best lobbyists—get priority.

Competition can work in public and private programs and force the system to be more responsive to consumers. By properly structuring incentives and creating a climate friendly to this innovation, Congress could put us on a path to uniquely American health care solutions.

IMPACT ON . . .
YOU AND YOUR DOCTOR

No matter how we reform health care, we will keep this
promise: If you like your doctor, you will be able to keep
your doctor. Period. If you like your health care plan,
you will be able to keep your health care plan. Period.
No one will take it away. No matter what.[1]

—President Barack Obama
June 15, 2009, speech to the American
Medical Association, Chicago, Illinois

The president's plan is now law. But even before most of it goes
into effect, we know those promises will not be kept.

As we explain elsewhere, as many as 80 million to 100 million people could have their insurance coverage changed by the time ObamaCare fully takes effect, and many will be shifted to different networks of doctors.[2] Some health insurers have stopped selling health insurance altogether,[3] forcing policyholders to find new plans and doctors. Total enrollment in Medicare Advantage plans will be cut in half, by as many as 7.4 million over the next ten years,[4] pushing seniors back into traditional Medicare, where fewer doctors are taking new patients.

So for millions of Americans, the prospects of keeping their doctors don't look good. Many physicians are considering leaving the practice of medicine altogether before ObamaCare fully takes effect.[5] Those who remain in practice risk becoming paid agents of federal health policy. The new law adds many new regulatory conditions governing the practice of medicine, including even more reporting requirements and government quality standards to be developed and enforced by federal officials.

Many doctors fear that ObamaCare will erode their professional autonomy, drown them in bureaucracy, and drain their job satisfaction. As the quality of their professional life continues to deteriorate, older doctors will retire while many younger doctors will look to switch careers. Many students considering a career in medicine are more likely to pursue other opportunities. The supply of doctors will dwindle as the demand for medical services reaches an all-time high. Ultimately, the consequences of the health overhaul law will be passed along to patients through restricted access, long waits for appointments, and even rationed care.

It wasn't supposed to be this way. The law was supposed to improve the practice of medicine. As we will explain, there are numerous parts of the legislation designed to increase the quality of care, to provide better coordinated care, and to reduce medical errors. But ObamaCare goes about this in the wrong way— through more government rules and regulations.

Health care is already one of the most highly regulated industries in the country, and doctors and nurses devote a significant amount of their day to detailed paperwork. This takes away from the time they could be spending with their patients. Reporting requirements will increase under the health overhaul law, and more physicians will be subject to financial penalties if they can't or won't comply with what could be an avalanche of new rules.

The "spirit" of ObamaCare is a spirit of centralization. While the Washington establishment will doubtlessly applaud the continued centralization of health care decision making—putting doctors on salary, experimenting with new "capitation" payment schemes, and herding them into larger group practices—independent and entrepreneurial physicians will see an unwelcome change in their practice of medicine. Economic and regulatory incentives will work against their initiative in ordering the tests, consultations, and medical procedures they think individual patients need. Many will be frustrated, even more than they are today, and with the growth in red tape, their ability to help and advocate for their patients will be weakened. Many will look for the exits.

The Patient Protection and Affordable Care Act (PPACA)[6] is indeed bad for doctors, but the new rules could dramatically affect your relationship with your doctor and even whether he or she can afford to continue to keep you as a patient. The new rules and regulatory burdens will also further discourage nurses—who also got into the profession to take care of patients, not paperwork.

"If you voted for Obama, seek urologic care elsewhere. Changes to your healthcare begin right now, not in four years."[7]

—Notice on Dr. Jack Cassell's Mount Dora, Fla., office

HEALTH CARE REFORM:
BAD FOR DOCTORS, AWFUL FOR PATIENTS

ObamaCare rolls the dice with the American health care system. A much better approach would have been to build directly upon its great strengths—its stunning capacity for innovation and its high-caliber training and professionalism of doctors and nurses— and focus on fixing its central shortcomings: the inequitable and inefficient financing that distorts private insurance markets and public programs, especially Medicare and Medicaid.

The law's 2,801 pages authorize mandates, directives, and regulations that will give bureaucrats unprecedented ability to intrude into the medical decision-making process. And as providers are required to do more, they are on track to be paid less. This is not the prescription for a bright future for the medical profession. A Thomson Reuters study released in January 2011 found that 78 percent of the 2,958 physicians surveyed said that the impact of the health overhaul law will be negative for doctors.[8] Only 8 percent said it would be positive. Responses indicate that these physicians are anxious about the future of health care and are concerned for their practices and their patients. Sixty-five percent said that they expect the quality of health care in America to deteriorate over the next five years.

WILL YOU BE ABLE TO KEEP YOUR DOCTOR?

Notwithstanding the American Medical Association's high-profile endorsement of the new law, there is deep discontent among physicians over ObamaCare. Shortly after it was passed, an April 2010 survey of physicians conducted by athenahealth and Sermo found that 79 percent of physicians were less optimistic about the future of medicine; 66 percent said they would consider dropping out of government health coverage programs; and 53 percent would consider opting out of participating in third-party insurance altogether.[9]

Their opposition to the new law has since hardened. In August 2010, the Physicians Foundation completed another major survey of doctors and found that 67 percent of them had a "somewhat" or "very" negative initial reaction to the new law. A total of 74 percent said they would take steps to change their medical practice over the next one to three years.

Sixty percent of these doctors said that the new law will force them to close or restrict certain categories of patients: 93 percent of them will stop seeing or restrict the number of Medicaid patients they see, and 87 percent said that they will close or restrict their Medicare practice.[10]

Ominously, 89 percent of physicians said that they believed that the survival of the traditional model of independent private medical practice is threatened. In fact, hospitals already own more than half of medical practices,[11] and that trend will be accelerated under the new health law. Will you have problems getting care under Medicare? If the law remains in force, the answer is almost certainly yes. Analysts at the Office of the Actuary of the Centers for Medicare & Medicaid Services (CMS), the powerful agency that runs these giant government health programs, say so. In an August 2010 report, they said that the number of hospitals, nursing homes, and hospice centers facing financial losses under the new law would jump to "roughly" 25 percent in 2030 and 40 percent by 2050.[12] Many Medicare providers will be forced to stop seeing Medicare patients, drop out of the program, or go bankrupt.

HEALTH CARE PLAN RULES
WILL RULE OVER DOCTORS

A key section of the health care law is called "Affordable Choices of Health Benefit Plans."[13] This says that states are required to establish exchanges in which customers can purchase health insurance. The law grants the secretary of health and human services significant authority to determine which plans will be allowed to participate as "qualified health plans." Though the effects of exchanges on insurance companies have received significant coverage in the media, their influence on physicians has received little attention.

This part of the law also gives the federal government new avenues for controlling physician practices. It calls for a payment system that rewards, among other things, "quality reporting . . . chronic disease management, medication and care compliance initiatives . . . use of best clinical practices, evidence based medicine, and health information technology." The secretary of HHS and her officers will write the rules that spell out in detail what this all means. And this is not voluntary: starting in 2015, a "qualified health plan" is allowed to contract with physicians and other medical providers "only if such provider implements such mechanisms to improve health care quality as the Secretary may by regulation require."[14]

The state exchanges create new federally supervised marketplaces for health insurance, called exchanges. Most health insurance companies will need to participate in the exchanges to stay in business. But the plans will be required to ensure that the doctors with whom they contract are practicing according to the new federal rules. This provision is an unprecedented expansion of federal regulation into contractual relationships between what are supposed to be private health plans and members of the medical profession. This extends the power of federal regulators far be-yond their current jurisdiction over Medicare and Medicaid today.

Who, at the end of the day, should determine what health care we receive? For most Americans, the answer is patients and doc-tors, not distant political officials, including the HHS secretary, Kathleen Sebelius. The tacit assumption behind the effort to have the federal government be the final arbiter about what is or is not quality care, and how and what medical treatments should be cov-ered or reimbursed, is that the government knows best, not you and your doctor. All third-party payers, including private insur-ance, interfere with the doctor-patient relationship, and we want less of this, not more. One of the goals of free-market reforms is to empower patients and lessen that interference as much as pos-sible. ObamaCare goes in precisely the opposite direction.

Doctors don't treat an "average patient"; rather, they treat the patient sitting before them. The new law, however, in an effort to improve quality care, provides new levers that the government can pull to move doctors toward standard government-approved practice patterns. This regimented approach could restrict access to clinical innovations and limit patients' choices. Many patients could rightfully be upset.

For doctors, this will be very demoralizing and frustrating, es-pecially since they will be witnessing the adverse consequences firsthand. Depending on the regulations, physicians may have in-centives to refer their patients to specialists other than the ones they have been seeing. Or to prescribe different drugs from the ones the patients have been taking. Or to explain why they are de-laying tests and not prescribing some treatments at all. Inevitably, patients' conditions could deteriorate because the government's "best clinical practices" are not best for them.

Perhaps the biggest tragedy is that these changes could hurt those that need the most help: the sickest, poorest, uninsured, and most troubled and noncompliant patients. If the government eventually implements rigid guidelines and ties doctors' fees to its outcome measurements, doctors will have a greater incentive to see patients who are responsible and compliant. Doctors could lose out financially when patients do not take care of themselves or don't follow treatment instructions. This could deter them from wanting to care for patients who have greater health needs but present greater challenges.

Even if they deeply resent the new regulatory regime, physicians may find there is little else to do but to comply. The "third-party payment" system has become ingrained in the medical community. Physicians' only option to avoid this loss of autonomy would be to opt out of Medicare, Medicaid, and most federally regulated insurance entirely. And some will; count on it.

Meanwhile, physicians in many states continue to labor under outdated medical liability laws that make malpractice insurance prohibitively expensive. They know that the new law will make matters even worse. An accumulation of new requirements, ranging from new Medicare rules governing patient referrals and reporting, plus tougher civil monetary penalties and criminal sanctions, will make the practice of medicine even riskier.

MORE GOVERNMENT PAYMENT AND RULES

As noted, 16 million of the 32 million people who would gain coverage during the first ten years would be enrolled in Medicaid, a welfare program jointly administered and funded by the federal government and the states. In 2013, the law prescribes that primary care physicians participating in Medicaid will receive Medicaid payments of no less than 100 percent of the Medicare payment rates for two years, 2013 and 2014. For these short-term

Medicaid payment increases, the new law authorizes federal taxpayers to pick up 100 percent of the additional funding. But once federal taxpayer funding ends, it will be up to the cash-strapped states to pay the doctors.

But most physicians' payments in the major entitlement programs, particularly Medicare and Medicaid, are generally less than the rates paid by private insurance. As a result, some seniors have trouble finding a doctor who will take new Medicare patients. And for Medicaid patients, the access problem is worse, and the care is all too often substandard, particularly for serious cases. For example, in 2009, 192,370 American women were diagnosed with breast cancer, and 40,170 died of the dreadful disease. While the mammography screening rate for breast cancer was 71.3 percent for women with commercial insurance, it was only 52.4 percent for those under Medicaid.[15]

COMPLY OR ELSE

The new law toughens the existing Physician Quality Reporting Initiative (PQRI) in Medicare, a fixture of the current law designed to improve the quality of care for Medicare beneficiaries. Supporters of ObamaCare contend that a more centralized approach to medicine could both coordinate care and serve patients better. This theory, once again, is based on the faulty premise that federal regulators are to be trusted more than independent medical professionals.

The basic assumption is that federal officials will properly define and competently apply quality standards. It works like this. If your doctor complies and reports the required data to the federal government, he gets a bonus payment of 1 percent of his Medicare payment in 2011. For the next three years, if he complies, he gets a

0.5 percent bonus payment for each year. In submitting information on the quality of care, the doctor has to be diligent: " . . . if the eligible professional does not satisfactorily submit data on quality measures for covered professional services for the quality reporting for the year," he or she will face a financial penalty.[16] Until 2015, the Medicare doctor's participation in the PQRI program will be voluntary. But in 2015, participation will become mandatory. Doctors who don't comply will face a 1.5 percent Medicare payment penalty, rising to 2 percent in 2016 and beyond.

Compliance with federal standards for medical practice is hardly a prescription for professional independence or clinical innovation in the delivery of care. But it does guarantee that medical professionals will have to spend more time proving they are adhering to rapidly multiplying Medicare rules. Physicians and their office staffs already devote a significant amount of their day to detailed documentation, exhaustive paperwork, and frequent signatures. These requirements will surely increase, at the expense of their time with patients.

Another mechanism for constraining doctors' professional independence is the newly established Value-Based Payment Modifier.[17] This will adjust the Medicare physician reimbursement formula to include cost and quality measurements, as defined by the secretary of HHS. By 2017, this new formula adjustment will establish a threshold for acceptable physician costs, and doctors whose costs are above the threshold will be penalized financially.

Doctors want to practice medicine to prescribe what is right for their patients, but their ability to do that very likely will be compromised by ObamaCare. This can be personally frustrating for physicians who want to help their patients, and it could also create new tensions as patients get upset when physicians are reluctant to recommend appropriate tests, medicines, or other interventions. The restrictions are a double-edged sword. PPACA does not protect doctors from lawsuits. Thus, the government will punish doctors for doing too much, but if a doctor doesn't do everything and a patient experiences a bad outcome, he or she also could be exposed to lawsuits.

Applying the Medicare value standards to hospital payments is likely to lead to administrators looking over physicians' shoulders even more closely than they do today. Currently, oversight is focused primarily on decisions to admit and discharge patients. The new payment rules mean they will have to explain their practice patterns, including CAT scan orders; MRI orders; and other tests, procedures, and consults. This will be incredibly draining for physicians who want to provide the best care for their patients. Furthermore, the implementation and enforcement of these compliance measures designed to save money will, ironically, be quite costly. It will also waste physicians' valuable time and energy. They will have to spend even more of their day in front of executives instead of with patients. These new mandates could easily be the final straw for some. Doctors already have a host of reasons not to accept Medicare patients, and this will be one more to add to the list.

HARMING THE DOCTOR-PATIENT RELATIONSHIP

As reimbursements inevitably decline, doctors will have little choice but to see more patients in less time. Doctors' daily schedules are already full, and patients' visits are already very short. It remains to be seen how these encounters can be made any shorter, but no doubt they will. Inevitably, the tighter time constraints of medical practice will significantly strain patients' relationships with their physicians. Time pressures will further push doctor-patient interactions away from meaningful conversations about treatment and compliance since doctors will have even less time to educate, counsel, answer questions, and offer explanations to their patients. As a result, patients will be less likely to understand their diseases and how best to treat them. This will understandably be frustrating for patients and could impact their compliance with the doctor's recommendations.

Another troubling feature of the new law is its insufficient protection for the rights of conscience of doctors and other medical professionals relating to the sensitive issue of abortion. The new law provides only limited protection of the rights of conscience for doctors and other medical personnel who refrain from participating in abortions.[18] The Bush administration published regulations in 2008 to enforce conscience protections by withholding federal funds from state or local entitities that violated providers' conscience rights and authorized the Office of Civil Rights at HHS to investigate violations. The Obama administration rescinded the Bush regulations. While the first version of health care reform legislation that passed the House in 2009 provided for robust conscience protections, the final law did not include this conscience protection language.

EXACERBATING THE PHYSICIAN SHORTAGE

Even before ObamaCare, we were faced with a severe physician shortage. There simply are not enough doctors, doctors in training, nurses, and nurse educators. In 2005, the Council on Graduate Medical Education issued a report on the status of the physician workforce. It concluded that even though the absolute number of physicians will expand between 2000 and 2020, the demand for them will grow at a relatively faster rate than the supply. Thus, the number of physicians per 1,000 patients will decrease.[19] In November 2008, the Association of American Medical Colleges (AAMC) examined this issue and predicted a national shortage of 124,000 full-time equivalent physicians by 2025.[20] While these deficiencies may be most significant in both number and newsworthiness in the field of primary care, the dearth of doctors actually encompasses most medical and surgical subspecialties.

Unfortunately, ObamaCare will only exacerbate this trend. After the passage of the legislation, the AAMC readdressed the supply of physicians and now envisions a shortage of about 160,000 doctors by 2025. This includes 46,000 primary care physicians and 41,000 general surgeons.[21] The new law will exacerbate the physician shortage. It offers insurance to the uninsured without significantly increasing the number of providers.

The law's National Health Care Workforce Commission is charged with studying the health care workforce and making recommendations to Congress and the executive branch for solutions. But a closer examination of these "workforce" provisions shows they largely replay the failed policies of the past and invite the intrusion of special interest politics.[22]

DECREASING THE SUPPLY OF PHYSICIANS

The health care overhaul law will worsen the physician shortage by decreasing physician supply. The onslaught of increased bureaucracy, additional paperwork, more oversight, and less autonomy will likely drain job satisfaction and could push some doctors to retire or switch careers. Some observers might postulate that the time doctors must invest before they can actually practice medicine would make their supply quite inelastic. The published literature, although limited in this area, suggests otherwise. Doctors, like any other employee, can and will tolerate only so much.

Researchers writing in *The Journal of Law and Economics* showed that practice climate does indeed influence physicians' work patterns. Examining physicians' work schedules in relation to malpractice risk, the researchers determined that doctors worked 1.7 hours less per week when the risk of medical liability increased by 10 percent. This effect was most pronounced for older physicians and physicians who own their practice.[23] Malpractice and regulations are certainly different issues. Nonetheless, just as the fear of malpractice drains job satisfaction and lowers take-home pay, so too do costly, time-consuming government regulations.

Physicians agree. A January 2010 survey of nearly 1,200 doctors conducted by Medicus, a national physician search firm, found that almost one-third of physicians and nearly half of primary care physicians would want to leave medicine if health care reform passed.[24] Obviously, not every doctor who said he or she would quit will, but when faced with more hassle and red tape, many will seriously entertain this option.

DETERRING YOUNG PEOPLE FROM PURSUING MEDICINE AS A PROFESSION

ObamaCare may also dissuade young people interested in medicine from pursuing it. This will likely be health care reform's main influence on the physician shortage. A career in medicine requires a tremendous investment in both time and money.[25] By the time doctors start their career, most are in their mid-thirties and have accumulated an average of $150,000 or more in education-related debt.[26]

Residency is not only mentally challenging but physically and psychologically demanding. Its rigorous training consists of regular thirty-hour shifts and eighty-hour workweeks. When residents work these lengthy shifts on a floor block, they typically get only four days off per month. Throughout this training marathon, they receive an hourly compensation, which hovers just above minimum wage. This salary lacks any correlation with responsibility, skill set, or education. For young physicians, medical training is not just a job; it's their lives. Becoming a physician requires major sacrifices. Doctors should be rewarded for giving up so much for this noble calling. They certainly should not be further discouraged or punished.

ObamaCare will deter bright young minds from entering medicine altogether. The best and the brightest are more likely to go elsewhere.

Fewer new physicians, coupled with an exodus of those already in practice, will create an access nightmare for patients desperately seeking care. While it will be a nuisance for doctors to have to change practice style or switch careers, ultimately the patient will suffer the most. Many new patients will not be able to find a physician, and in time, despite President Obama's "if you like your doctor" promise, many patients who are content with their health care and their doctor will have no choice but to change.

TROUBLE GETTING CARE UNDER MEDICARE

When the Medicare program was created, government was not supposed to dictate in any way how your doctor practiced medicine.

The original Medicare law, passed in 1965, says that federal officials are not to exert any control over how medicine is practiced or how medical professionals are paid. Under Section 1801 of Title XVIII, Congress set clear boundaries for federal officials:

> Nothing in this Title shall be construed to authorize any federal officer or employee to exercise any supervision or control over the practice of medicine or the manner in which medical services are provided, or over the selection, tenure or compensation of any officer or employee of any institution, agency, or person, providing health services; or to exercise any supervision or control over the administration or operation of any such institution, agency, or person.

Seems pretty clear. But forty-five years—and more than 100,000 pages of rules, regulations, and guidelines—later, a lot has changed.

Your protection from government interference in the doctor-patient relationship has been seriously eroded in Medicare,[27] and it's going to get worse.

There are well over a hundred sections in ObamaCare dealing with Medicare, ranging from changes in payments for hospitals, skilled nursing homes, and home health care agencies to major new rules and requirements that will change how your doctor practices medicine.

According to analysts at the nonpartisan Congressional Research Service (CRS), the agency that prepares reports for Congress, these changes will have a far-ranging impact on the practice of medicine. The CRS says the law can "affect physicians and how they practice in ways both small and large, immediately and over time."[28]

By altering the terms and conditions of your doctor's practice of medicine, as the CRS observes, the new law will indeed change your future relationship with your doctor.

We have described only a few of the new provisions here. There are many more, such as the new Accountable Care Organizations designed to improve the quality of care, which we describe in more detail in the chapter "Impact on . . . Seniors." But one of the ways that ObamaCare will most dramatically affect Medicare patients is that so much money will be taken out of the program to pay for new entitlements. And the main tool government will use to do this will be to pay providers less.

THE UNSOLVED MEDICARE PROBLEM

During the health reform debate, supporters of the health overhaul law praised Medicare's ability to exploit its size to obtain lower fees with providers. While it is true that Medicare can bludgeon down physicians' fees, this is not one of the program's greatest strengths. In fact, it is one of its greatest weaknesses. Those underpayments are ultimately shifted to patients in the form of shorter visits, less face time with their doctors, quick discharges from hospitals, and a growing unwillingness among physicians to take new Medicare patients.

At first glance, this is the doctor's problem and not that of the patient. Unfortunately for the nation's seniors, this is not the case. Ultimately, the consequences are passed along to the patient. The most obvious is difficulty finding a physician. This problem is increasing. The American Academy of Family Physicians discovered that 13 percent of doctors surveyed did not participate in Medicare in 2009, up significantly from 6 percent in 2004.[29] The harmful influence of the current policy is broader and more complicated. In the outpatient setting, low reimbursements force doctors to see more and more patients to stay in business. This leads to shorter visits and less doctor-patient face time. It forces doctors to race through medical appointments so they can fit as many patients into the daily schedule as possible, frantically bouncing from exam room to exam room.

These shorter visits are not only frustrating but dangerous. Researchers writing in the *British Medical Journal* studied consultation length and linked shorter medical visits with the inability of patients to understand and cope with illness.[30]

Once again, centralized decisions over payment and policy often disregard the needs of individual patients.

PATIENT CENTERED OUTCOMES RESEARCH

Section 6301 of PPACA establishes the Patient Centered Outcomes Research Institute (PCORI), a nonprofit, federally funded organization, which will identify research priorities and conduct research comparing the efficacy of medical and surgical interventions. PCORI will disseminate this research to health care providers and patients. The legislation creates a Board of Governors to run PCORI and a trust fund to support it financially.[31]

Supporters of PCORI believe this research could educate patients and help them make better medical decisions. Some physicians will welcome the government's treatment protocols. This organization indeed *could* foster patients' education and benefit both patients and doctors. In fact, researchers creating hypotheses, conducting research, challenging perceived norms, and publishing in peer-reviewed journals is modern medicine at its finest. The potential harm from PCORI depends on how the research is used or abused. It also could quell medical innovation by sub-stituting budgetary for medical decisions or by making political decisions under the guise of science.

The real and legitimate fear is that officials seeking to control costs would use this research to restrict access to more costly medical interventions, as is done by Britain's National Institute for Health and Clinical Excellence (NICE). As promised savings from new government programs fail to materialize, there will be tremendous pressure on elected officials to slow spiraling costs. Writing in *The Cancer Journal*, Dr. Stuart M. Butler of The Heritage Foundation observes, "Given the administrative state approach to health sector management and cost control, however, it seems very likely that the PCORI will to some degree evolve into the role occupied by NICE. The reason is that when the ap-proach is to regulate centrally the availability of, and payment for, health services based on comparative effectiveness, there has to be an expert body that either makes the determinations or recom-mends determinations to federal agencies."[32]

The Congressional Budget Office told Congress that to signifi-cantly decrease spending using these tools under ObamaCare, the government would probably need "additional legislative author-ity to allow the program to consider relative benefits and costs in a more extensive way and to modify the financial incentives facing doctors and enrollees accordingly."[33]

President Obama has promised that comparative effectiveness research won't lead to rationing. His problem is that his recess appointment of Dr. Donald Berwick to head CMS and implement Medicare payment changes has raised the rationing issue, without providing Dr. Berwick the well-deserved opportunity to appear at a Senate confirmation hearing and explain his views on the issue. The record shows that Dr. Berwick has expressed lavish praise for NICE. In a 2009 interview with *Biotechnology Health Care*, he said that NICE has "developed very good and very disciplined, scientifically grounded, policy-connected models for the evaluation of medical treatments from which we ought to learn."[34] But the agency is notoriously controversial.

Ultimately, there are only two ways to lower costs. One approach empowers and incentivizes patients to be smarter health care consumers. This entails solutions such as providing premium support, creating a national market for health insurance, and leveling the tax playing field for individuals and health plans to compete in an information-driven economic environment. Such changes could bend the cost curve down by engaging market forces that would also have the effect of strengthen the doctor-patient relationship.

ObamaCare reflects a radically different approach. It is based on centralization of health care decision making in the hands of government officials, the use of administrative payment formulas for provider reimbursement, and the exaltation of the interests of society as superior to the welfare of the individual patient.

A one-size-fits-all model for health care does not benefit patients. Patients are individuals, not programmed robots. They have a wide range of personal treatment preferences. When presenting with back pain, some patients want the most invasive procedures available despite potential significant side effects, while others prefer more conservative treatment. Is a patient suffering from back pain wrong to prefer over-the-counter aspirin and rehabilitation? Is another patient with back pain wrong to want a riskier laminectomy? Who knows best? Perhaps they both do. Physicians should work with patients to determine the best treatment course. A regimented approach to care could ignore such individual preferences.

Mandating the same treatment regimens and algorithms for every patient seems to violate basic individual rights and freedoms. It also contradicts current medical trends toward more individualized and personalized care. This is another crucial reason that we must get on a very different path to health reform.

CONTROLS ON FUTURE MEDICARE PAYMENT

Physicians are trained to bring their knowledge, skills, and abilities to taking care of individual patients. Under the Hippocratic Oath, the ethical tradition that has governed the profession for approximately 2,500 years, physicians are cast as the servants of their patients. Nothing should come between them and their patients. But today, many medical schools don't even administer the traditional oath anymore or they administer only a watered-down version.[35]

In Medicare, Medicaid, and private insurance, this traditional doctor-patient relationship has already been compromised by putting payment administrators between patients and their doctors. If you're enrolled in Medicare or Medicaid, federal officials set the prices and conditions for how your doctors are paid for taking care of you. You never know or meet the government officials who make those decisions. If you don't like their decisions for some reason, you may take the time to write your congressman, in the hope that he or she knows something about the payment rules Congress enacted. Chances are, though, that you'll get a very unenlightening form letter—ostensibly from your legislator but most likely written by a young staffer, telling you that he "shares your concerns." That's nice to know, but it doesn't help you much.

The decline of traditional medical ethics has largely taken

place under the radar; for the most part, this has been a quiet revolution. Nonetheless, it will profoundly affect the delivery of health care, particularly at the beginning and the end of life.

ObamaCare doubles down on this quiet revolution. It will drown doctors in red tape and bureaucracy. It will limit physicians' autonomy and their ability to help and advocate for their patients. Their job satisfaction will be drained, and the doctor-patient relationship will be seriously compromised. As federal regulators require physicians to do more, they are likely to be paid even less. As the situation worsens, older doctors will retire and younger doctors will look to switch careers. This will come at a time when the demand for physicians' services will be higher than ever. Ultimately, ObamaCare will translate into restricted access to and inferior quality of care. No matter how you look at it, this legislation is terrible for physicians; however, it is always the patient who suffers the most.

The only way to restore the balance is by empowering patients and providing new incentives for them to become smarter health care consumers. As we describe in our chapter "What We Should Do Instead to Get Reform Right," this involves providing a defined contribution for health care benefits, creating a national market for health insurance, and leveling the tax playing field. This is the way to get costs under control, strengthen the doctor-patient relationship, and provide incentives to usher in a new generation of modern medical miracles for the twenty-first century.

IMPACT ON . . .
YOU AND YOUR EMPLOYER

Why do most people get their health insurance at work? It's a question we seldom ask because it's just assumed that is what we do here in the United States.

Workers see health insurance as a valuable job benefit that provides them and their families with stable coverage. And many employers see it as a good investment that contributes to a healthy workforce. It provides the foundation for our private health insurance market.

Today, about 170 million Americans get health insurance through the workplace.[1] Any social institution that covers that many people and has lasted for well over half a century must be doing something right. It certainly should be entitled to the benefit of the doubt in any health care reform proposals.

But the system is not working today for the millions of people who have lost their jobs. Or who can't get work. Or who can't afford their share of the increasingly expensive premiums. Or whose employer simply can't afford to offer insurance. The system needs updating to help the millions of people who are being left behind. They need other options to get stable, affordable health insurance (as we describe in our chapter "What We Should Do Instead to Get Reform Right").

In light of its dominance in the market, it's worthwhile to look at the advantages—and disadvantages—of job-based insurance. After that, we will describe how ObamaCare is going to change it.

As we will explain, your insurance at work is under siege and in danger. ObamaCare first plans to harness and redirect your job-based coverage to serve its own purposes. But the ultimate goal of ObamaCare is really to replace it in whole or in part with even more government-centered health insurance.

ADVANTAGES OF EMPLOYER COVERAGE

Dating back to World War II, the government has provided strong financial incentives to get health insurance at work.[2]

Because of federal tax policy, the premiums that an employer pays for its employees' health insurance don't count as taxable income. This is called a "tax exclusion." It's different from a "deduction," where we write a check and then deduct it from our taxes—such as tax deductions for charitable contributions or mortgage interest. With the tax exclusion, taxpayers and the IRS never see the money. The money for health insurance is taken off the top, or excluded, from employees' income before their employer calculates the taxes they owe.[3]

It's cheaper to buy health insurance with pretax dollars than to buy it on our own with after-tax dollars.[4] The tax savings can be worth several thousand dollars a year. So over the decades, consciously or not, tax policy has provided a big financial incentive for tens of millions of Americans and families to get health insurance at work.

But there are other advantages as well. Most workers get and keep their jobs for reasons other than health insurance. As a result, large companies, especially those with a diverse group of employees, generally provide a stable risk pool. This allows insurers to more accurately estimate the group's likely health care costs.

It also provides economies of scale. When people buy insurance on their own, insurers have to do more expensive and intrusive underwriting to predict each individual's health risks. With a large employer group, the insurer is able to estimate the expected cost of health care claims based upon the group's claims in the previous years. The law of large numbers works to employees' advantage. Eliminating individual underwriting lowers overhead costs. And insurers save on marketing costs when they can offer a limited number of policy options and sell them to an employer group as a whole rather than to each person separately.

Large employers have additional advantages in sponsoring health insurance plans. They can bargain more effectively as bulk purchasers for lower insurance premiums and discounted health care provider fees. They usually have the administrative resources to assist their employees in searching for, organizing, managing, and simplifying information about options for health benefits, health care providers, and medical treatments. Many employers also help their employees appeal inappropriate denial of claims by insurers and negotiate better prices for their health benefits.

Employers don't offer health coverage just because of special tax advantages or out of benevolence. Health benefits help them to recruit and retain the workers they need and maintain or improve their workers' health. That means less absenteeism and turnover, lower disability and sick leave costs, and improved productivity.

Finally, private employers have flexibility to experiment with benefit plans.[5] For example, employers have offered wellness and weight loss programs coupled with incentive payments as a way to engage employees as partners in better managing their own health and health spending.[6] These innovations have helped to keep employer health costs lower than they otherwise would have been. But this flexibility will be threatened as employers and workers are forced to fit into ObamaCare's centralized insurance standards.

THE LIMITS OF EMPLOYER COVERAGE

Though there are advantages, there are also serious limitations to our system of employer-based coverage. Employees have the mistaken impression that their boss is paying for most, if not all, of their coverage. That leads them to demand more coverage without realizing that they are paying for these more expensive health benefits out of their total compensation package. More expensive health insurance means less take-home pay.

Smaller firms also don't enjoy many of the savings of larger firms in offering health insurance. Their risk pools are smaller and less stable, they have fewer economies of scale, and one sick employee can send costs for the whole group soaring. Small employers may also be able to offer only one insurance plan.

Employees can "keep what they have" only as long as they keep their current job. And then there is the major problem that changing jobs usually means changing health insurance—or losing it altogether.[7]

There are other people who are completely shut out of these coverage advantages who wonder why they are treated differently. If they can afford insurance, they generally have to pay the full after-tax price, and they face the extra costs and constraints of state insurance regulation of fully insured plans.

The solution is to equalize tax policy so everyone gets the same tax break, whether you get your health insurance on your own, through other groups, or at work—as millions no doubt would continue to do in the reformed system we envision and describe in the last chapter.

But that's not what ObamaCare does.

ONLY A SHORT-TERM TRUCE WITH EMPLOYER-BASED COVERAGE

President Obama and his aides repeatedly assured employers that they didn't have anything to worry about with his reform plan; their plans would be "grandfathered" in

Behind the scenes, the White House needed to keep employers in the game so they would continue to finance the great majority of private health insurance. ObamaCare relies greatly on maintaining the outer shell of most employer coverage while controlling what is inside through new mandates, regulations, cross subsidies, and political bullying. Moreover, it lacked sufficient political support to replace employer coverage more openly with a "public option" coverage road to a single-payer health care system.

But after the law was enacted, the Obama administration invented new rules for this "grandfathering" protection so that it was much harder for employers to comply. To be protected from a barrage of costly new mandates, an employer has to fit into a straitjacket, limiting the changes it can make to its existing plans. Something as simple as changing a copayment could toss an employer out of the protected "grandfather" category. At that point, the employer would be forced to comply with a number of costly new ObamaCare rules.[8]

Employers must weigh any advantages from making changes to their plans that would cause them to lose their grandfathered status—but lower their rising premiums to some extent—against the increased burdens of the additional PPACA mandates. In other words, they will have to choose their poison (even higher premiums versus new regulatory costs).

This is the first big double-cross for employers.

A survey by the health benefits consulting firm Mercer found that only about half of the 1,100 companies it polled believe they are likely to retain grandfathered status for all of their plans in 2011.[9] Federal regulators assume that by 2013, only about one in five small employers and one-third of large employers will remain grandfathered.[10]

So the early rounds of the health overhaul law impose a number of new regulatory burdens and related costs on most employer health plans.[11] These will push their premiums higher, rather than fulfilling President Obama's 2008 campaign promise of lower health care costs.[12]

Over the longer term, the law sets into motion incentives, mandates, and taxes that will push employers away from continuing to offer health coverage and pull many of their workers toward getting their coverage through the new health benefits exchanges, where they can get taxpayer-subsidized insurance.

In other words, ObamaCare may be hoping for the employer goose to keep laying its golden eggs, but most of its main policies squeeze it too hard and are more likely to cook it extra crispy.

REALITY BITES IN 2014

These initial mandates and regulatory restrictions are mere "appetizers" for other requirements to come. The full force of ObamaCare will come in January 2014. That's when workers will have to start proving that they have government-required health insurance and employers will have to prove that they are either "playing or paying" for their workers' insurance.

If you work for or own a company with fifty or more employees, your company will be required to offer government-defined "minimum essential coverage" health insurance that is "affordable" and pays at least 60 percent of covered claims costs, or it must pay fines.[13] A few companies will manage to escape through "grandfathering," but all health insurers will be restricted to offering to employers or individuals one or more of four government-approved tiers of coverage (along with a catastrophic plan for young adults). Employers must offer the policies they choose to all applicants regardless of health status and without excluding preexisting conditions.[14]

And while the law doesn't require small employers with fewer than fifty workers to provide insurance, it does regulate the premiums that can be charged in those small groups.[15]

EXIT SIGNS ARE FLASHING FOR FUTURE EMPLOYER-BASED COVERAGE

ObamaCare's new, enhanced requirements for employer health plans, coupled with the lure of new subsidies for state-based health benefits exchanges, could spell the end of the health insurance that millions of workers have today. The law sets into motion several strong incentives for companies to drop or substantially change their existing employer-based coverage, as well as change whether they hire new workers and how they pay them.

One expert with McKinsey & Company, a management consulting firm, predicted that in late 2010 "something in the range of 80 to 100 million individuals are going to change coverage categories in the two years post-2014."[16] Alissa Meade told those attending a Chicago-based conference on health insurance exchanges that the new health care reform law will bring "fundamental disruption to the health care economy."

Meade and McKinsey also are predicting that 30 to 40 million people will choose to remain uninsured by 2016. They will refuse to buy expensive coverage under the individual mandate and will decide to pay the fine instead. ObamaCare could make it so difficult to comply with the law that millions of people actively decide to defy it. What kind of a civil society will that create?

"To avoid additional costs and regulations, employers may consider exiting the employer health market and send employees to the Exchanges." [17]

—Document prepared for Verizon by Hewitt Resources

UNINTENDED CONSEQUENCE: BARRIERS TO JOB AND WAGE GROWTH

As part of its mandate that employers must provide "qualified" health insurance beginning in 2014, ObamaCare attempts a misguided carrot-and-stick approach that will discourage small firms from growing and that will especially discourage them from hiring lower-wage workers.

The first thing ObamaCare will do is discourage firms from crossing the fifty-employee threshold, at which point the full employer mandate and its enforcement penalties begin to kick in. The basic penalty is $2,000 for each full-time (or equivalent part-time) employee.[18] Although smaller firms are "exempt" from these penalties, the effect will be to discourage them from growing and hiring more workers lest they hit the magic 50 mark.

Larger employers most likely will continue to provide insurance and escape the $2,000-per-worker penalty. They will, however, face another threat: they will have to pay $3,000 for each of their employees who turns down employer coverage when their share of premiums is too high relative to their income (or when the employer plan pays less than 60 percent of covered insurance claims) and gains other subsidized coverage in a state-based health benefits exchange.

Here it gets really tricky. The new law says that workers should not have to pay more than 9.5 percent of their household income for health insurance. The key word here is "household." How is an employer supposed to know how much the spouse or even working children of their employees make? This makes it most risky and costly for employers to hire workers who earn the least but who probably need the job the most. They are the ones most likely to trigger the $3,000 penalty.

This also opens doors to privacy questions. If your employer is going to be fined if he doesn't know your household income—and every time it changes—your boss has a strong incentive to learn a lot more about you and your family life than you may want him to know.

If you are married, how is your employer going to know how much your husband or wife makes? What if you get separated or divorced and your household income changes? Your employer will need to know that, too, in order to make sure it isn't violating federal law. And the government is going to need to know your income and employment status *every month* if you are eligible for a subsidy in the exchanges. If your income changes and you receive a larger subsidy than your income level warrants, you will have to pay back some or all of the premium subsidy.

Tracy Watts, a partner with Mercer, is concerned: "Lawmakers did not take into account that employers don't have access to information on employee household income. Employers question how they are going to get that information and what other administrative challenges might come along with this new requirement. For example, what happens if an employee's total family income changes during the course of a plan year?"[19]

Never fear. We're sure that ObamaCare's regulators are hard at work on new rules to figure this out.

LITTLE RELIEF FOR SMALL BUSINESSES

The new health care law pretends to provide substantial assistance to small businesses that offer health insurance to their workers. But there are a lot of strings attached to the small-business tax credit. The actual subsidies will turn out to be small and temporary (for a maximum of six years). They are also linked to limits on future wage and job growth in any small businesses that initially receive them.

The credits are available only to employers with no more than twenty-five employees. The full subsidies begin to phase out for employers with eleven to twenty-five workers and for firms with average annual wages per employee of $25,000 to $50,000.

In short, only firms offering qualified insurance coverage are eligible for the tax credits, and hiring too many workers, or paying them higher wages, will reduce or eliminate a small firm's eligibility. Not surprisingly, the Congressional Budget Office estimates that the small-business tax credit provision would benefit only 12 percent of small businesses.

Is this any way to grow an economy, when all the signals from Washington say "Slow down" or "Stop" to small employers hoping to get bigger and to workers hoping to move up the wage ladder? The small-business tax credits are structured to *raise* the bar on small-business expansion and wage growth, which will discourage business improvements and investments.

NEW TEMPTATIONS FOR EMPLOYERS TO DROP COVERAGE

On top of all of these mandates, fines, and regulatory burdens, ObamaCare adds another headache for employers: new taxes and paperwork requirements.

Higher payroll taxes . . . new taxes on insurance premiums, on medical devices and on prescription drugs that will drive up the cost of health insurance . . . requirements to track and report to the IRS all purchases from vendors totaling more than $600 each year.[20]

Where will it end?

Companies are having a hard enough time making a profit and keeping their businesses afloat. ObamaCare could sink more of them. According to a report issued by the IRS' national taxpayer advocate, 40 million businesses will be affected by the health care law's new 1099 paperwork mandate.[21] Another study estimates that the publicly reported costs of early compliance with the new health law to corporate employers totaled more than $4 billion as of November 2010.[22] Few of these new burdens in and of themselves will determine whether employers continue to offer health insurance. But they amount to a steady, drip-by-drip water torture that can eventually reach critical mass and push employers to drop health insurance altogether.

Former football coach Bill Parcells once said, "They want you to cook the dinner; at least they should let you shop for the groceries."[23] In the case of ObamaCare, employers are increasingly being asked first to pay for health insurance groceries selected by Washington regulators and then to cook and serve them according to recipes concocted in Congress and at HHS.

THE LURE TO LEAVE

The final blow to "keeping the coverage you have" will come from the deceptive lure of highly subsidized coverage that will be offered through the new health insurance exchanges, starting in 2014. This insurance will seem like a great deal to many workers—especially lower-wage workers, whose premiums will be heavily subsidized by taxpayers.

However, these generous subsidies are officially limited to families earning between 100 and 400 percent of the federal poverty level ($22,050 to $88,200 in 2010 dollars) *who do not receive qualified health insurance from their employers or from public programs such as Medicaid.*

There's the catch. You can get this subsidized insurance only if your employer doesn't offer federally approved insurance.

So let's look at the incentives here: Your employer will be fined $2,000 per employee if it doesn't offer coverage. But the coverage may cost your employer several times that amount—at least. It could make more sense for it to just drop your coverage, pay the fine, and send you to the exchange. No employer wants to be the first, but once it happens, there will be a race for the exits.

ObamaCare tries to create a "firewall" to prevent workers from leaving their plans at work and opting for the subsidized coverage in the exchanges. It assumes that the mandate for companies with fifty or more employees to provide insurance, plus penalties for violating it, will keep them from dropping insurance. The new health care law further assumes that the mandate on individual workers to obtain qualified coverage will force them to accept whatever plan their employer offers them.

But what happens if those firewalls become politically flammable and this house of cards for subsidized coverage begins to go up in the smoke of even more massive taxpayer liabilities? The tax breaks that low- to middle-wage workers get today for employer-sponsored insurance (through the tax exclusion) are much less valuable than the subsidies they would receive if they opt for coverage in the exchanges. This state of affairs is neither financially nor politically sustainable.[24]

Former Congressional Budget Office director Douglas Holtz-Eakin estimates that employers will respond very differently to the exchange subsidies than ObamaCare's architects assumed. Many employers will be much better off financially if they drop coverage for employees, pay the employer penalty, and increase their workers' wages as an offset. Their then-uninsured employees would seek the highly subsidized coverage in the new exchanges instead.

This will be true particularly if most of a company's employees make less than $88,200 a year (the level at which subsidies in the exchanges end). Holtz-Eakin and his coauthor Cameron Smith estimate that there may be as many as 35 million more workers by 2014 for whom it makes sense to drop employer-based insurance. That figure would nearly triple the total number of Americans (to 54 million) that the Congressional Budget Office assumed last year (19 million) would make use of the new exchange coverage subsidies. That would send the price tag for taxpayers soaring for just this one part of ObamaCare. This program to provide subsidies for people to get private insurance was expected to cost as much as $450 billion over 10 years. But Holtz-Eakin expects it to reach $1.4 trillion![25] That's one budget estimate that was not "close enough for government work."

"While we clearly need health-care reform, the last thing our country needs is a massive new health-care entitlement that will create hundreds of billions of dollars of new unfunded deficits and move us much closer to a government takeover of our health-care system."[26]

—John Mackey, Co–Chief Executive Officer, Whole Foods

Those massive taxpayer liabilities by themselves are unsustainable. But add to this bill of unintended consequences other items such as the accompanying disruption and reworking of existing compensation packages, insurance coverage arrangements, and labor market relations. The overall toll will be devastating throughout the overall economy.

Further, it is likely to lead to a two-tier system of health insurance where workers in high-wage firms, such as law and accounting firms, keep their generous private health insurance while middle- and lower-income workers will be relegated to the exchanges, where coverage will begin to resemble that under Medicaid: doctors will be paid less and less as costs rise, it will be harder and harder to get appointments, and new technologies will be rationed as the government tries in vain to contain spiraling costs.

ADDING TO EMPLOYERS' UNCERTAINTY

ObamaCare continues to produce massive uncertainty and confusion for employers and their workers. "What will they do next?" is a question that will needlessly distract and divert businesspeople from the more important economic challenges they face as our economy slowly recovers from the deep recession. The sheer difficulty of understanding, anticipating, and maneuvering through the changing and complex regulatory terrain of ObamaCare will be difficult enough for large firms. For smaller firms struggling to survive and hoping later to thrive, the new health care law essentially tells them to hire fewer workers, keep wages down, and reduce investment in their operations. ObamaCare will lead to slower wage growth, fewer job opportunities, and more businesses going under. Moreover, this news is bad not just for your employer and other businesses; it will be translated, to a lesser degree, into higher prices when you go shopping.

But "at least you will have your health." Maybe. Don't count on that either, at least from ObamaCare.

IMPACT ON . . . TAXPAYERS

P resident Obama and his allies in Congress insisted throughout the health reform debate that their plan not only would cover all Americans with health insurance but also would reduce the federal government's massive budget deficits. And they said they would do this by cutting the growth of health care spending.[1]

Americans were skeptical. How could it be possible to provide health insurance to 32 million or more people without increasing costs, raising taxes, or adding to the deficit?

The president also promised that Americans who already have health insurance would see their premiums fall, so that the average family would save $2,500 a year.

Americans were right to be skeptical. ObamaCare won't cut costs for any of us, nor will it shrink the government's budget deficits or bring down the debt. The president's claims are based on accounting gimmicks, old-fashioned smoke and mirrors, and deceptive advertising.

It's clear that ObamaCare is a budget buster of historic proportions. It is the largest expansion of government in nearly half a century. It will add trillions of dollars to government spending in the coming years, and it will force even higher taxes on working Americans in the future.

No, ObamaCare is not the solution to our budget problems. It is a budgetary disaster. ObamaCare resorted to every trick in the book to pretend that it won't add "one dime" to the deficit. It raises taxes for ten years but counts spending for only six years. It pretends to spend the same dollars twice—once on budget savings and once on creating new entitlements. It relies on spending cuts that will never happen. It raises hundreds of billions of dollars through new job-killing taxes. And it seriously underestimates the number of people who will receive huge subsidies for its expensive health insurance.

SPENDING AND TAX INCREASES

Even with smoke and mirrors, ObamaCare will mean massive spending and taxes.

The president and his allies say that the new law will cut the government's budget deficit by $124 billion over the next ten years and by another $1 trillion in the decade after that.[2] That's based on an assessment produced by the Congressional Budget Office.[3]

But even this optimistic projection reveals that ObamaCare is a massive tax and spending bill.

Government spending will go up by about $1 trillion (that's $1,000,000,000,000) over ten years, according to the CBO, and taxes will increase by more than $500 billion ($500,000,000,000) over the same period.[4]

And the tax increases won't hit just "rich" people. Obama-Care will raise more than $100 billion in new taxes over the next ten years on insurers, drug manufacturers, and companies that make medical devices. Those taxes will quickly be passed on to middle-class consumers in the form of higher health insurance premiums.[5]

In addition, there are new add-ons to the Medicare payroll taxes (0.9 percent on wages and 3.8 percent on incomes from other sources). Initially, those new taxes will affect only families with incomes above $250,000 a year ($200,000 for single people). But the thresholds don't rise with inflation. So as the years pass, more and more families will find themselves paying these taxes as their incomes rise.[6]

That's exactly what ObamaCare's defenders are hoping will happen, because they have created a new entitlement monster that must be fed with more and more revenue every year.

HUGE NEW SPENDING

ObamaCare will add about 16 million people to the 58 million already on the Medicaid program. And it will pay all or part of the health insurance premiums for another 19 million Americans whose family income is under $88,200 and who buy insurance through the new state-run health exchanges.[7]

Even with the rosiest of assumptions, CBO expects these new and added programs to increase government spending by $210 billion in 2019.[8] And once they start paying benefits, their cost will soar, rising faster than wages or revenue every year.

These are the numbers that ObamaCare's apologists cite when they are trying to reassure taxpayers about what they have done. It doesn't generally work. Most Americans are worried about the government piling $1 trillion in new spending and $500 billion in new taxes on the already overburdened economy.

But as bad as they are, even these numbers don't begin to tell the real story. ObamaCare is going to push government spending up even more than these best-case-scenario numbers show, and the result will be to force even higher taxes on future generations of Americans. That becomes clear only when all of the gimmicks and unlikely assumptions that are hiding ObamaCare's true costs are stripped away.

"There is a very significant chance this is going to be a nightmare." [9]

—Indiana Governor Mitch Daniels at an
American Enterprise Institute conference

SIX YEARS OF NEW SPENDING PAID FOR WITH TEN YEARS OF CUTS AND TAXES

In August 2009, tens of thousands of Americans went to congressional town hall meetings to protest the $1 trillion health care bill Congress was then considering. They were shocked at the price tag on the president's plan to reorganize U.S. health care.

But that didn't slow the president down. He pushed ahead despite the opposition. His only concession was a pledge, made in a speech before Congress in September, to hold the overall costs of the effort to "only" $900 billion over a decade. [10] The new entitlements that ObamaCare creates will cost much, much more than this $90 billion a year. In fact, the cost will be more than $200 billion a year by 2019.

So how did the president and his allies squeeze a $2 trillion program into a $900 billion budget? For starters, they didn't.

To even get to $1 trillion, ObamaCare's sponsors resorted to the simplest gimmick of all: they decided to start the tax increases and spending cuts in 2010 but not start the entitlement spending until 2014.

That's right. The president argued it was a necessity to enact his health care plan in 2010 to cover the uninsured. But to create the perception that it will cost "only" $1 trillion, he makes the uninsured wait nearly four years before they get any help. But the tax increases have already started.

Once the spending starts, costs will soar. The costs of the ObamaCare entitlement will reach about $2.3 trillion over the first ten years of full implementation.[11]

THE MEDICARE DOUBLE COUNT

ObamaCare relies heavily on cuts to the Medicare program to pay for new health care benefits for working-age people. The CBO says the cuts in Medicare spending will total about $450 billion over a decade.[12] The Obama administration claims that these cuts will replenish the Medicare trust fund and that they will make Medicare stronger so it can pay benefits in the future.[13]

But the same dollar can't be spent twice, no matter how hard Congress tries.

Although most Americans don't need confirmation for what they know from common sense, both the CBO and the experts who project Medicare's costs for the president offered their confirmation anyway. They went out of their way to let it be known that the Medicare cuts can't be double-counted. Either they will pay for a new entitlement, or they will shore up Medicare.[14] But not both.[15]

THE CLASS ACT GIMMICK

ObamaCare also creates a new program for long-term care insurance—the Community Living Assistance Services and Supports program, or CLASS.

The program is voluntary, and people who are eligible would be required to pay premiums long before they might receive any benefit payments.[16] Though that isn't surprising for a program like this, its main role in ObamaCare was to provide another budget gimmick to make it look as though the law was "paid for."

Because CLASS premiums will be collected for five years before the first dollar is paid out in benefits, it appears that Obama-Care will cost less than it does. But at some point, after the first five years, CLASS insurance claims also will need to be paid. So where is the government going to get the money to do that? Through higher taxes or more deficit spending.

The CLASS Act is a piggy bank rigged to explode. Obama-Care's authors again could not resist the temptation to double dip. They set aside the premiums to pay future long-term care insurance claims, but they also counted the premiums to pay for ObamaCare's overall insurance expansions. They've counted $70 billion twice—once to save for future benefit payments under the CLASS Act and another to pay for the new entitlements.

When this double count is removed from the budget accounting (along with the Medicare double count), the added debt burden from ObamaCare moves toward $300 billion over ten years.

But that's not all that's wrong with the CLASS Act. Senate Budget Committee Chairman Kent Conrad (D-N.D.) called CLASS "a Ponzi scheme of the first order, the kind of thing that Bernie Madoff would have been proud of."[17] It's easy to see why. The program is very poorly designed. It is voluntary, which means it will draw enrollment mainly from those who think they are likely to need long-term care services. The premiums will be high, which will discourage healthy people from signing up, but not high enough to cover all of the program's costs.

So not only does ObamaCare count the premiums for CLASS twice, it also underfunds it. Future taxpayers will almost certainly be asked to bail out this new program when the premiums that are paid by participants don't cover the full costs of the long-term care insurance claims that have to be paid.[18] So this program is triple jeopardy for taxpayers.

THE "DOC FIX" AND OTHER UNREALISTIC MEDICARE CUTS AND TAX INCREASES

Every year since 2002, Congress has stepped in to undo automatic cuts to what Medicare pays physicians. The cuts are built into the federal budget process as a way of holding down federal spending, and they happen automatically unless Congress takes action to delay them. Both Republicans and Democrats in Congress have voted to override the cuts, and for good reason: they are arbitrary and irrational and would cause large numbers of physicians to stop seeing Medicare patients.

Unfortunately, Congress has yet to find a way to permanently undo the cuts because there has never been enough money around to pay for the $200 billion to $300 billion over ten years that is needed to fix the problem. So it has passed a series of one-year "doc fixes" to push the cuts forward in time. The cuts are now scheduled to occur in 2012, and, if this happens, it will slash what Medicare pays doctors by nearly one-third.[19]

Everyone knows that these cuts shouldn't happen. The only question is how to fix the problem permanently.

The version of ObamaCare that was introduced in the House

of Representatives in 2009 actually did fix the problem permanently. The leaders did this to get the American Medical Association (AMA) to endorse their bill.

But then the president gave his speech to a joint session of Congress, pledging to keep the cost of his health care plan to "only" $900 billion over a decade. The original House version of Obama-Care overshot that mark by about $300 billion (which is, not coincidentally, just about the cost of a permanent doc fix).

So enterprising Democrats settled on a solution to get the cost of their bill down to within shouting range of the president's goal. They decided to pass the "doc fix" separately from ObamaCare. Not only that, they decided to stop trying to find ways to pay for the "doc fix." They would just add it to the national debt.

In other words, they hoped to hide the cost of the "doc fix" by passing it separately from ObamaCare.

The only problem with this plan was that even some Democrats in the Senate found it shameless and refused to support it, undercutting the House leaders' efforts to pass it. But this resistance didn't force House Democrats to put a permanent fix for Medicare physicians back onto their priority list; they simply dropped the "doc fix" idea altogether.

ObamaCare did nothing to address the problem. Physicians still face deep cuts in what Medicare pays them, putting at risk seniors' access to care. The cost of fixing this is still $300 billion or more, with no money left after ObamaCare to do anything about it.

The amazing thing is that the AMA still endorsed Obama-Care! It was told that someday in the future, Congress would fix the problem. And its endorsement was crucial to ObamaCare's passing. Not only did Congress not fix the problem, it made it much more difficult ever to find a solution.

HOSPITAL COSTS TARGETED

ObamaCare also targets hospitals and other entities providing services to seniors by making cuts that are deep, unrealistic, and damaging.[20] The authors did this to create the illusion of keeping ObamaCare's cost under $1 trillion. Congress is using these supposed "savings" to grease the way for its massive new entitlement.[21] But, just like the cuts for doctors, the cuts to hospitals and others won't hold up.

Normally, hospitals and other institutions providing care to seniors receive a payment increase every year to cover the rise in their own costs. Under ObamaCare, those inflation increases will be cut by about half of a percentage point every year, forever.

Doing this every year will, over time, dramatically reduce what Medicare pays. Medicare's chief actuary, Richard S. Foster, has stated plainly that these cuts will not stand up over time. By the year 2020, Medicare payments would be below what *Medicaid* pays, even though Medicaid's rates are so low that many hospitals and doctors can't even cover their costs when they treat Medicaid patients.

ObamaCare's apologists would like Americans to believe they have set into motion a sophisticated and carefully considered plan to slow spending in Medicare. The truth is that all they have done is put into law a formula that calls for deeper Medicare price cuts. But these payment cuts always have the same predictable effect. Fewer doctors will be able to afford to see Medicare patients. Patients will have to wait longer to get an appointment. Waiting lists will grow. It will be the beginning of rationed care.

This is part of a long-term pattern. It's the way the government has always handled health care cost control. To hit budget targets, Congress and Medicare's regulators simply cut what they pay to doctors, hospitals, and others treating Medicare patients. Obama-Care is no different.

The strategy has never worked to control costs before and won't this time either. Doctors and hospitals make up for lost fees by doing more each time they see a patient. They order more tests and do more procedures. And Congress later overrides many of the cuts by adding back payments (and usually increasing the national debt).

GET READY FOR HIGHER TAXES

So the cuts in what the government spends on health care won't really slow spending. ObamaCare's tax increases are equally implausible.

If you paid taxes in the 1970s, you might remember the Carter-era "bracket creep." That's what happens when the income tax brackets don't keep up with inflation. As wages rise each year, more and more people are pushed into higher tax brackets. They pay more in taxes even when their real income stays the same.

The Reagan tax cuts ended this "bracket creep" for income taxes. Now ObamaCare is bringing it back for payroll taxes.

Initially, it will hit individual taxpayers with annual incomes exceeding $200,000 and couples with incomes exceeding $250,000. They will have to pay an additional 0.9 percent in Medicare taxes on wages and 3.8 percent on nonwage income. But those income thresholds will stay the same and not increase over time. Consequently, normal inflation will mean that, over the years, more and more Americans, including the middle class, will pay those higher taxes.

Overall, the Medicare tax hikes are expected to raise about

$1.4 trillion over seventy-five years (discounted for inflation). But that assumes that America's middle class will placidly accept the return to the bad old days of "bracket creep." A more realistic assumption is that elected leaders will come under pressure very soon to prevent such a massive tax hike on the middle class. Congress will likely respond, much as it does today with the alternative minimum tax, by delaying the tax hikes.[22] And every year it does, more and more deficit spending will be needed to finance ObamaCare.

TAXES ON HIGH-COST INSURANCE

ObamaCare does something similar with what has become known as the "Cadillac insurance plan tax." This is the new excise tax on expensive insurance plans—hence the name. Candidate Obama adamantly opposed taxing health care benefits during the 2008 presidential campaign and in fact ran millions of dollars of attack ads against Senator John McCain on this issue.

President Obama, however, changed his mind once he got into office and called for a tax on expensive health plans.

But this flip-flop infuriated his organized labor allies. They argued that this kind of tax punished them because their members tend to be older than the average worker, and the plans they have negotiated are more generous—and expensive. Therefore, firms with more union workers are more likely to cross the threshold and face the Cadillac insurance plan tax.

Under pressure, President Obama made a last-minute concession. He would delay the tax until 2018. But once it was in place, he would allow the threshold to increase at the rate of general inflation, but not the faster inflation rate of health costs. That means that in 2018, very few people would be in plans that are taxed. But in the years after that, as health costs rise faster than inflation, many millions of people would pay the tax.

So, again, here is an example of implausible and deceptive budgeting. President Obama didn't have the political courage to collect the Cadillac insurance plan tax himself. By the time it starts in 2018, he will have left office, even after a possible second term. But he wants us to believe that the next president and future Congresses will have the political courage to collect the tax. Not only that, he wants us to believe they will collect it from millions and millions of middle-class people who will find themselves in plans that exceed the limit—something he himself was not willing to do.

THE COMING ENTITLEMENT EXPLOSION

ObamaCare is built on the dubious assumption that Congress can hand out a lucrative new benefit to a limited number of citizens while denying the same benefits to tens of millions of other people.

The centerpiece of ObamaCare is a promise of new subsidies for health care to all households with incomes between 133 and 400 percent of the federal poverty line. For a family of four, four times the poverty line was about $88,200 in 2010.[23]

ObamaCare promises to limit how much those families will have to pay for health insurance. For example, a family at 133 percent of the poverty line (about $29,327) would have to pay only 3 percent of its income for health insurance premiums. The federal government would pay the rest.

This will be a very large benefit provided by the government. Here's an example: health insurance for a family of four is expected to cost about $14,100 in 2016.[24] Under ObamaCare, a family of four with an income of about $44,000 in 2016 will pay only about $2,700 toward this insurance, as long as they buy it through one of the state-run insurance exchanges that ObamaCare creates.[25] The rest of the premium—more than $11,000 in this example—will be paid by federal taxpayers.

Once again, it is a shell game. ObamaCare pretends that it can lower health insurance premiums, but it is doing so with money that has to come from taxpayers or by adding to the deficit.

FOLLOW THE MONEY

The number of people potentially eligible for this new federal entitlement is huge. The Census Bureau says there are about 111 million Americans under the age of sixty-five with incomes in that range.[26]

But ObamaCare's sponsors didn't want to provide assistance to all those millions of people. That would send costs soaring. So they wrote a rule into the law that makes anyone who is offered qualified insurance by his or her employer ineligible for the extra federal assistance.

That's how the Congressional Budget Office is able to assume that only 19 million people—of the 111 million who are eligible—will receive the new federal health insurance subsidies in 2019.[27] The CBO expects that most employers will continue to offer qualified insurance to their workers. That means that most workers, even if their income is low enough to qualify for the new government assistance, will not be eligible for it.

Of course, insurance offered by an employer is also subsidized by the government because health care benefits are not taxed like wages. But the value of this tax preference is worth much less than ObamaCare's new health benefits. For a family of four with $60,000 in wages in 2016, the new entitlement in the exchanges would be worth $3,500 more than the tax break they receive for buying health insurance at work.[28]

Employers with large numbers of low- and moderate-wage workers are going to figure out that they and their workers would both be better off if the workers got their health insurance through the state-run exchanges and not on the job because of the added governmental assistance. As new businesses are formed, they could organize, in part, to take advantage of both the subsidies available in the exchanges and the tax break that remains for those with higher incomes.

So instead of just 19 million people receiving subsidized insurance from the government, the number is likely to be much higher. Former CBO director Douglas Holtz-Eakin estimates that 35 million more workers and their families with incomes below 250 percent of the poverty line would be better off in the exchanges, where they could get more generous government subsidies. When they migrate to the exchanges, it will send the cost of ObamaCare soaring. Instead of it costing $1 trillion between 2010 and 2020, it will cost about $2 trillion[29]—or more!

"Our federal deficit is already at unsustainable levels, and most Americans understand that we can ill afford another entitlement program that adds substantially to it."[30]

—Former Tennessee governor Philip Bredesen

MORE SPENDING PILED ON HUGE DEFICITS

In its first two years, the Obama administration embarked on an unprecedented spending binge. Total government spending in the two-year period 2009–2010 exceeded what had been spent in the previous two-year period by $1.3 trillion.[31] The federal government is piling up new debt at rates not seen since World War II. If the Obama budget plan were adopted in full, the government would run a deficit of $10 trillion over the coming decade.[32]

What's needed is a serious plan to cut government spending.

But ObamaCare will go in exactly the wrong direction, by piling more spending obligations on top of the unaffordable ones already on the books.

Meanwhile, to pretend to pay for it all, ObamaCare's supporters resorted to every trick in the budgetary book: delayed starts for the spending but not for the taxes . . . double counting of money for two entitlements . . . Medicare cuts that will never actually happen . . . and tax increases that start small but balloon over time.

It's dead certain that the spending will expand even more, as entitlements always do, coupled with flimsy or nonexistent ways to pay for it. Federal entitlement programs always grow and expand. The new entitlement created by ObamaCare will be no exception. It will cost much more than projected.

At some point, the real bill for ObamaCare will come due. And when it does, Americans will be proven right: ObamaCare's promises were too good to be true.

IMPACT ON . . . YOU AND YOUR CONSTITUTIONAL RIGHTS

ObamaCare's operating system and underlying values are fundamentally at odds with the principles of limited government under the U.S. Constitution. The new health overhaul law is designed to increase the dependence of citizens on political decision makers in Washington. It places government in the driver's seat to determine many personal health care matters. Patients, providers, and employers in the private health care sector must scramble to meet the ever-changing demands of unaccountable regulators. When citizens and companies are not begging for waivers and special deals, they will be waiting to receive further instructions.

ObamaCare imposes an unprecedented mandate on individuals to purchase government-approved health insurance. It mandates that states expand their Medicaid programs and operate prefabricated health benefits exchanges, thus reducing state governments to little more than branch offices of Washington regulators. All of this rearrangement of the careful constitutional balance between the rights of citizens and the responsibilities of their elected representatives was cemented in the new health law by a series of unsavory backroom political deals that have further undermined the public's faith in Congress.

But this will not stand. Outraged Americans first began to protest spontaneously during town hall meetings in the summer of 2009 as the initial legislative version of ObamaCare took form. After it was enacted, private citizens and organizations, as well as many state government officials, went to federal courts to defend their constitutional rights. This series of lawsuits is primarily directed at the individual mandate and the limits of federal power. More than half (twenty-six) of all state governments have joined one of these cases, and two others have filed separate suits on their own. Two states have approved new amendments to their state constitutions to protect their constituents against future excesses of ObamaCare. Others are exploring potential compacts among the states to reset and rebalance the terms of their future participation in federal health care programs.

THE "DEPENDENCY" AGENDA

The nationally syndicated columnist George F. Will framed this issue best shortly before the new health law was finally enacted in March 2010:

> America's dynamism, and hence upward social mobility, will slow, as the economy becomes what the party of government wants it to be—increasingly dependent on government-created demand.
>
> Promoting dependency is the Democratic Party's vocation. Democrats believe, plausibly, that middle-class entitlements are instantly addictive and, because there is no known detoxifica-

tion, that class, when facing future choices between trimming entitlements or increasing taxes, will choose the latter. The taxes will disproportionately burden high earners, thereby tightening the noose of society's dependency on government for investments and job creation.[1]

We don't have to cast this in purely partisan terms, as Will did. Although support for ObamaCare in Congress was very much limited to Democrats, key Republican leaders and officeholders have also been tempted in the not-too-distant past to play the dependency card. (For example, the unfunded Medicare prescription drug benefit was initiated by President George W. Bush in 2003. Earmarks for pork-barrel spending exploded out of control in a Republican-controlled Congress throughout the first half of the last decade. And the switch to automatic cost-of-living increases for Social Security beneficiaries was signed into law by President Richard M. Nixon in July 1972—which was, not coincidentally, an election year.) The larger strategy in health care over the last two to three years, as outlined by Will, has been to promise the implausible—prompt and noticeable improvements in the increasingly complex and annoying health care system. Public complaints about the failure of new rules, mandates, and taxpayer subsidies to deliver the promised results will inevitably be "smothered by more subsidies," he predicts.[2]

A FORK IN THE ROAD TOWARD HEALTH CARE SERFDOM?

Is this dependency cycle as automatic and self-reinforcing as this? Although government funds have begun to account for more than one of every two dollars spent on health care, the private sector has not yet been crowded out completely. But the ambitions and appetites of ObamaCare's designers reach well beyond the economics of gaining a larger market share for government-directed health care. If and when it is fully implemented, Obama-Care will transform the relationship between citizens and their government and reverse their roles of master and servant. You will be required by a legal mandate to buy government-approved health insurance. You will face the Hobson's choice of either being more economically and politically dependent on government officials to help you purchase this expensive brand of insurance or becoming an increasingly overtaxed source of subsidies for others. More and more of the key decisions regarding some of the most personal health care issues you face will be determined outright (or narrowly limited) by unelected and largely unaccountable government regulators and bureaucrats.

All of this increased dependency on government-centered preferences and interests will spill over into the demanding and unpredictable politics of seeking and obtaining health care. With the personal becoming so political, citizens will bargain as subordinates for the best deal they can get rather than instruct office-holders to reflect their principles and preferences. "Tell us what we have to do" will replace "Represent us faithfully" in voters' messages to Washington-based politicians. A "virtual" federal takeover of health care does not have to be absolute to become sweeping and entrenched. Retaining the outward appearance of privately operated health care institutions and health insurance carriers will mean little if they have to operate more like heavily regulated public utilities and if potential new competitors are shut out of health care markets.

UNACCOUNTABLE AND ARBITRARY
OVERSEERS OF YOUR HEALTH CARE

The new health law tried to deflect initial concerns about a full-fledged federal takeover of private health care by relying on ambiguity, deferred decision making, and administrative complexity, all buried in thousands of pages of legislative text.[3] Even after Congress first asserted authority over nearly everything imaginable in health care, it had to delegate a vast amount of work to interpret and implement it to hordes of current executive branch agencies and new boards, commissions, and offices. The key political objective was to gain permanent regulatory authority over a vast expanse of health care operations and decisions—regardless of how many blanks needed to be filled in later. By continuing the appearance of private operation of most current health care arrangements—but under much tighter (and potentially expanding) political control—the hope was that this pending takeover of the health care sector could be partially disguised.[4]

However, it will be hard to keep this vast expansion of federal power over health care under wraps for long since it is hiding in plain sight. The legislation will weave new webs of bureaucratic red tape that will lead to more plea bargaining by special interests. The regulatory machine already is grinding away. (So many commands to issue, words to redefine, schemes to concoct, favors to grant, and lives to control. And so little time!)

SHORTCUTS AND SPEEDUPS

The Congressional Research Service (CRS), which provides objective research to Congress on the law and on the history of federal public policy, noted that most of ObamaCare's complex and lengthy rules thus far were not specifically required by the legislation. And most of them went into effect without giving the public much, if any, chance to comment ahead of time.[5] Obama administration officials, mostly in the Department of Health and Human Services, frequently relied on the device of an "interim final rule" to speed them along. They didn't tell people the rule was coming or give them time to send in comments. They just issued the rule. Clearly, any new information or criticisms from citizens or companies about how they would be affected by the rule were unwelcome. The regulators are in a hurry to cement into place the first pillars of a new, far-reaching command-and-control regime for health care. The CRS observed that the short comment periods allowed in most cases would likely limit citizens' input and concluded that after-the-fact "comments on final rules are generally believed to be less likely to result in changes to the rules than if comments were permitted prior to the final rules being published and made effective."[6]

BENDING, EXTENDING, OR DODGING THE "RULES"

Of course, not all of the rules are equal in the eyes of the Obama administration, and neither are the rules for when and how they are issued. Broad discretionary authority tempts (and often allows) federal officials to interpret vague statutory language and terms more broadly or narrowly. They can also make exceptions to the rules they issue, grant waivers, or apply them differently. For example, the PPACA simply states, in Section 1251, that its "Immediate Improvements in Health Care Coverage for All Americans" and "Quality Health Insurance Coverage for All" do *not apply* to group health plans and health insurance that exist as of the date the legislation was enacted: March 23, 2010. However, this clear and direct "grandfathering" protection of existing insurance arrangements was quickly transformed by ObamaCare regulators into a more complex and overreaching final rule for grandfathered health plans that made such protection highly unlikely for them![7] On the other hand, Obama administration regulators arbitrarily decided to delay another requirement of the new health overhaul law, that of early reporting by employers of the cost of the insurance benefits they provide to employees, despite the explicit date of calendar year 2011 provided in the statute.[8]

WHO DESERVES A BREAK TODAY?

Between these two extremes of upgraded and downgraded statutory limits was the initial reaction to the administration's efforts to extend new rules for "annual benefits caps" for insurance plans to so-called mini-med limited-benefits plans, offered mostly by smaller employers with high-turnover, short-tenure workers. The disconnect between the new rules' "one-size-fits-none" standards and the real-world efforts by companies to stretch employees' compensation during tough economic times came to light vividly in the fall of 2010. The McDonald's Corporation warned that nearly 30,000 of its hourly restaurant workers might end up a few mini-med plans short of a "Happy Meal" if regulators did not waive the new requirements for them.[9]

Facing the politically untenable headline of "ObamaCare Kills Jobs of Low-Wage Workers," Health and Human Services Secretary Kathleen Sebelius quickly pivoted. She essentially told McDonald's executives "You deserve a break today"—just as the 2010 midterm elections were approaching. She granted the high-profile corporation a special, temporary waiver from rules that dictated what its policies must cover and what services it could afford to provide. Within months, more than 700 other employers, insurers, and unions sponsoring health plans were granted similar waivers. But as the window was closing for seeking such short-term dispensations from ObamaCare's edicts (the equivalent of a few more hours of exercise time in the prison yard), only the quick, clever, and politically connected portion of the employer community would be able to get a waiver.[10]

The only operating rule for the new crew of health care regulators appears to be "With great power comes great irresponsibility, random inconsistency, and selective enforcement"—along with intensified lobbying for special treatment, backroom deal making, and redirected unfairness. Just like the process in Congress that led to the law getting passed. The only certainties are that special-interest lobbying on health care regulation will explode and that bureaucratic politics will determine more and more of your personal life.

MANDATES VERSUS THE CONSTITUTION

The glue that will try to bind all of these rules and regulations together involves an unprecedented mandate on individuals to purchase government-approved health insurance. The clamps are the onerous mandates on states to expand their Medicaid programs and the requirement that they set up and operate complex new bureaucracies called health benefits exchanges that satisfy the marching orders given by the secretary of HHS.

PRESIDENT OBAMA AND THE
INDIVIDUAL MANDATE

Voters might remember that presidential candidate Obama never actually supported an individual mandate before he was elected, in contrast to his main opponent for the Democratic Party nomination, then-senator Hillary Clinton. On the campaign trail, he insisted that the important objective was to make health insurance affordable, which he said would make a mandate to purchase insurance unnecessary. He criticized supporters of mandates, including Mrs. Clinton, not only for trying to "force" people to buy insurance that was not affordable but for wanting to make them either pay for it out of their paychecks or pay government fines.[11] During one of the presidential candidate debates, candidate Obama pointed out, "If a mandate was a solution, we could try that to solve homelessness by mandating everyone buy a house."[12]

However, shortly after Election Day in November 2008, President-elect Obama at first quietly signaled that he was, after all, quite supportive of the wishes of almost all congressional Democrats to push an individual mandate as an essential part of the legislation that produced ObamaCare. The president apparently felt the need to move beyond his ambiguous presidential campaign stance, in which he had promised universal coverage, but only if it first was made "affordable" and did not require "a very harsh, stiff penalty." By June 2009, President Obama publicly acknowledged that his thinking on the issue had "evolved" and he now could support a law mandating that individuals purchase health care coverage, with fines for those who do not. However, he stressed that there must be some kind of waiver for those who are simply unable to afford it (apparently even after new subsidies are approved by Congress).[13]

Subsequent bills passed first by the House in November 2009 and then by the Senate on Christmas Eve of that year, as well as the final version of the PPACA, included an individual mandate. It was indeed the centerpiece of ObamaCare's scheme to promise near-universal coverage, even as it fell far short. For a number of reasons (but mostly political ones), Congress delayed until 2014 the trigger for the individual mandate and the new subsidies to help people pay for coverage. If you fail to purchase the required health care plan by then, you will have to pay a penalty, which will be levied by the Internal Revenue Service. Various exceptions and exemptions were provided for those who could not find affordable coverage even with the assistance of new subsidies.

THE POLITICS OF THE INDIVIDUAL MANDATE

The political case for an individual insurance mandate was actually weak. It was built on false hopes, empty illusions, and larger agendas.[14] The biggest push for the individual mandate came from the ideological impulse of those who wrote the law to tell everyone else what to do because they believe they know better than we do what is good for us. Some people and businesses supported the individual mandate in the naive hope that if they could make someone else pay more for insurance, they'd get to pay less. A number of politicians wanted to substitute the mandate that we spend our private funds on insurance for the taxes Congress would otherwise find hard to impose to meet its illusory goal of universal coverage. In trying to put together a broader coalition in favor of expanded insurance coverage, ObamaCare supporters thought that at least some employers would support an individual mandate as a way to avoid a mandate on employers to provide insurance. Some hospitals and doctor groups hoped a mandate would help ensure that more of their patients have insurance to pay more of their bills. And some private insurance companies were willing to tolerate tighter political control over their operations if they were promised more revenue from involuntary customers.

Perhaps it should be no surprise that an individual mandate has the least support from those it is purported to help—people who don't enroll in public programs, don't have employer insurance, or don't purchase individual coverage. They either prefer to spend their money on other goods and services of greater value to them or are simply too hard-pressed financially to afford the health insurance generally available to them. Others wondered why, if buying some kind of health insurance that's offered to them is such a good idea, they should have to be forced by law to buy it. Moreover, unless a large number of new people can be coerced into paying more for health insurance than it will actually be worth to them, insurance mandates will create a perpetual conflict among escalating costs, limited resources, and the false guarantee of rich coverage.

Many Americans are troubled by Congress imposing a legal mandate on citizens to purchase something regardless of their wishes. They worry that imposing an individual mandate would actually operate as a gateway drug to even greater addiction to government control of health care. A mandate, in practice, requires more and more rules regarding exactly what it requires, how it's carried out, and who pays for it. Effective enforcement is even more questionable.

The clear alternative to a mandate would be to improve the value (both the cost and quality) of health care and therefore the insurance that helps finance it and invest in other, more effective ways to boost one's lifetime health. There are less onerous ways to get health insurance to more people, such as providing more targeted and equitable subsidies; charging higher premiums for those who delay, or fail to maintain, coverage; and offering better products that customers will purchase voluntarily.

MORE FUNDAMENTAL RIGHTS AND PRINCIPLES

But more fundamentally, many Americans feel that an individual mandate to purchase health insurance violates core principles of economic freedom, personal choice, and limited government under the U.S. Constitution. As the noted constitutional law scholar Randy Barnett of Georgetown University has observed, the only other mandates ever imposed on American citizens have been related to duties of citizenship: the military draft, jury duty, filing income tax returns, and responding to the census. He warned, "If economic mandates like this one are allowed, Americans will be demoted from citizens to subjects. They will have to obey any commands that Congress deems convenient to its regulation of interstate commerce. No more expensive tax credits and subsidies to raise taxes to pay for; Congress can just command you to buy its favored products. . . . Gone will be a federal government of limited and enumerated powers established by the Constitution and repeatedly affirmed by the Supreme Court."[15]

STATES CHALLENGE CONSTITUTIONALITY
OF THE INDIVIDUAL MANDATE

Accordingly, the constitutionality of the individual mandate was quickly challenged in a number of lawsuits that were filed across the country shortly after the health care law was enacted. The two most prominent ones were filed by the attorney general of

Virginia and by the State of Florida (with nineteen other states initially joining the latter case as original coplaintiffs) on March 23, 2010—almost immediately after the PPACA was signed into law.[16] Both lawsuits, as well as several similar ones filed before other courts, argued that the individual mandate in Section 1501 of the new health law was beyond the limits of the powers of Congress under the Commerce Clause of the U.S. Constitution. The important distinction to be drawn here was that, although Congress has broad power to regulate interstate commerce, in this case it was improperly attempting to regulate an individual's decision *not to engage in economic activity*, and therefore in any interstate commerce. In other words, the decision not to purchase a product, such as federally mandated health insurance, was "passive inactivity" and beyond the reach of federal powers under the Constitution.

The states also challenged other justifications by the Obama administration for power to impose an individual mandate, under either the federal government's regulatory authority under the Necessary and Proper Clause or its taxing and spending powers under the General Welfare Clause of the Constitution.

THE LEGAL CASE FOR THE INDIVIDUAL MANDATE

For more than half a century, the Supreme Court and lesser federal courts have interpreted the reach of the Commerce Clause to be very broad indeed, including the power to regulate activities that "substantially affect interstate commerce." Since the early years of the New Deal in the 1930s, Congress and the federal government have expanded their role in regulating the economy. In *Wickard v. Filburn* (1942),[17] the Supreme Court upheld the power of Congress to regulate the personal cultivation and consumption of wheat on a private farm. It reasoned that even consumption of wheat not offered for sale could reduce the amount of commercially produced wheat that was purchased and consumed nationally, thereby affecting interstate commerce. Since then, the Supreme Court has imposed no limits on the federal power to regulate commerce in matters involving economic activity. In a more recent examination of the boundaries of the Commerce Clause, the Supreme Court ruled in *Gonzalez v. Raich* (2005)[18] that growing and using marijuana for personal, medicinal purposes under California law still had a sufficient overall impact on interstate commerce to warrant regulation under the Commerce Clause.

The secretary of HHS and the Obama administration's lawyers argued that there is a rational basis for concluding that decisions by the uninsured not to participate in the health insurance market have a collective effect on interstate commerce that poses a threat to a national market. Therefore, they say, the individual mandate is a necessary and proper measure to ensure the success of the larger health reforms in the PPACA, claiming it is an "an integral part of the regulatory scheme" for ObamaCare's plan to increase insurance coverage and lower health care costs. The secretary also pointed to a series of legislative findings (in the same bill, by the same Congress that enacted ObamaCare) that a failure to regulate the decision to delay or forgo buying insurance—the decision to shift one's costs onto the larger health care system—would undermine the comprehensive regulatory regime.

U.S. District Judge Henry Hudson ruled in Virginia v. Sebelius *in December 2010 that the individual mandate "would invite unbridled exercise of federal police powers" and "is neither within the letter nor the spirit of the Constitution."* [19]

VIRGINIA V. SEBELIUS STRIKES DOWN THE MANDATE

Several other federal district court judges in the Eastern District of Michigan and in the Western District of Virginia swallowed this line of argument and dismissed similar legal challenges to ObamaCare's individual mandate.[20] But Judge Henry Hudson in the U.S. District Court, Eastern District of Virginia, saw things very differently. In December 2010, he ruled in *Virginia v. Sebelius* that the means adopted to enforce ObamaCare's individual mandate must not violate an independent constitutional prohibition, and the law must relate to the implementation of a constitutionally enumerated power of the federal government.

Judge Hudson cited another Supreme Court ruling in *United States v. Morrison* (2000) that said congressional findings, no matter how extensive, are not sufficient to enlarge the Commerce Clause powers of Congress.[21] He pointed to the Court decision in that case, as well as those in *Perez v. United States* (1971) and *United States v. Lopez* (1995), that limited those congressional powers to activities that are truly economic in nature and that have a demonstrable effect on interstate commerce.[22] Judge Hudson dismissed the argument that because everyone will eventually require health care at some point in their lifetime, requiring people to purchase insurance in advance is an activity that will inevitably affect interstate commerce. He found that definition of "economic" activity too broad, lacking logical limitation, and unsupported by Commerce Clause case law. Hence, without some type of self-initiated action by an individual to place him- or herself within the stream of commerce, Congress could not use its Commerce Clause powers to compel an individual to purchase a commodity such as health insurance in the private market. He ruled that the individual mandate was "neither within the letter nor the spirit of the Constitution."

A PENALTY, NOT A TAX

The Virginia district court opinion also dismissed an alternative argument by Secretary Sebelius that the individual mandate was actually an exercise of Congress's taxation power under the General Welfare Clause of the Constitution. That argument stretched the boundaries of credibility. For example, just take the word of Congress: the individual-mandate section in the PPACA flatly stated that it was a product of the Commerce Clause, not the General Welfare Clause. Congress also specifically labeled the payment imposed on those who failed to comply with the mandate a "penalty," not a tax.

Judge Hudson concluded that the legislative purpose of the individual mandate was not generation of revenue but rather purely regulation of what Congress perceived to be economic activity. He also noted the unequivocal and vigorous denials by the executive and legislative branches that the individual-mandate provision, while being considered and debated for final passage, was a "tax." Earlier versions of the bill in both the House and Senate had initially used the less popular term "tax," but the final version of the PPACA deliberately substituted the term "penalty" in its place. Hence, the penalty provision must be linked to an enumerated power under the Constitution other than the General Welfare Clause. But Judge Hudson could find no previous federal court decisions that had extended federal authority to regulate a person's decision *not* to purchase a product.

He concluded that the individual mandate in the new health care law would invite the unbridled exercise of federal "police powers" traditionally reserved to the states. Judge Hudson determined that this dispute over constitutional law was really about an individual's right to choose to participate in universal health insurance, rather than simply about crafting that scheme or regulating the business of insurance. Therefore, he ruled that the individual-mandate provision exceeds the constitutional boundaries of congressional power.

FLORIDA V. HHS

Another federal district court judge in the *State of Florida et al. v. United States Department of Health and Human Services et al.* case was equally skeptical of the constitutional validity of the individual mandate and any argument that its penalty for noncompliance was actually a tax. Judge Roger Vinson's initial decision on October 14, 2010, to deny a federal government motion to dismiss the states' lawsuit found that Congress did not intend for the penalty for failure to comply with the individual mandate to be regarded as a "tax." He noted that Congress had called it a penalty, rather than a tax, in the final version of the bill it approved, and that Congress had specifically changed the bill's language to do so. Congress also specifically relied on and identified its Commerce Clause power rather than its taxing power, eliminated traditional IRS enforcement methods for failure to pay the penalty ("tax"), and failed to identify in the legislation any revenue that would be raised from it.[23]

Judge Vinson also explained that members of Congress should not be allowed to avoid accountability by shifting the label from calling something a penalty to deciding it is a "tax" so it can pass constitutional muster afterward. "[T]he members of Congress would have reaped a political advantage by calling it and treating it as a penalty while the Act was being debated . . . and [would] then reap a legal advantage by calling it a tax in court once it passed into law," he wrote, adding "Congress should not be permitted to secure and cast politically difficult votes on controversial legislation by deliberately calling something one thing, after which the defenders of that legislation take an 'Alice-in-Wonderland' tack and argue in court that Congress really meant something else entirely, thereby circumventing the safeguard that exists to keep their broad power in check."[24]

Accordingly, Judge Vinson ruled that the individual mandate penalty was not a "tax" and the federal government defendants may not rely on the taxing authority of Congress under the General Welfare Clause to try to justify the penalty after the fact. The penalty could be sustained only if it was imposed to aid and assist an enumerated power—such as the Commerce Clause power.

"Because the individual mandate is unconstitutional and not severable, the entire Act must be declared void."

—U.S. District Court Judge Roger Vinson, January 31, 2011

A GIANT EXPANSION OF THE COMMERCE CLAUSE

Regarding the claim that the Commerce Clause power allowed the federal government to impose an individual mandate, Judge Vinson summarized that argument as essentially saying:

(1) because the majority of people will at some point in their lives need and consume healthcare services, and (2) because some of the people are unwilling or unable to pay for those services, (3) Congress may regulate everyone and require that everyone have specific, federally-approved insurance.[25]

Judge Vinson emphasized that the Commerce Clause and the Necessary and Proper Clause of the Constitution have "never been applied in such a manner before."[26] He distinguished this case from older commerce power cases that involved activities in which someone had chosen to engage from this case of an individual mandate that applies across the board:

> Those who fall under the individual mandate either comply with it, or they are penalized. It is not based on activity that they make the choice to undertake. Rather, it is based solely on citizenship and on being alive.[27]

At that stage of the court proceedings in the *Florida v. HHS* case, Judge Vinson denied the federal government's motion to dismiss the states' claim that the line between constitutional and extraconstitutional government had been crossed in two respects (more on the second one below). In a subsequent hearing on December 16, 2010, on a motion for summary judgment in the case, Judge Vinson underscored his conclusion, remarking that "it would be a giant leap for the Supreme Court to say that a decision to buy or not to buy is tantamount to an activity. That would be a giant expansion of the commerce clause."[28]

He added, "In the broadest sense every decision we make is economic. The decision to marry. The decision to keep a job or not has an economic effect. If [the federal government] decided everybody needs to eat broccoli because broccoli makes us healthy, they could mandate that everybody has to eat broccoli each week?"

From tea party to broccoli party, anyone?

As this book went to press, Judge Vinson issued, on January 31, 2011, his final ruling in *Florida v. HHS*. He not only declared the individual mandate unconstitutional (elaborating on his reasoning in earlier rulings), his opinion struck down the entire health law as void because its other provisions were not "severable" from the unconstitutional individual mandate.

COMMANDEERING AND COERCING THE STATES

Unlike other lawsuits challenging the individual mandate, the twenty-six states in the Florida case (six more states joined as co-plantiffs in early 2011) also objected to the fundamental changes in the nature and scope of their Medicaid programs required by the new health care law. In expanding Medicaid to an extent that they could no longer afford, the new federal mandates would improperly "coerce and commandeer" them in violation of the Ninth and Tenth Amendments of the Constitution,[29] said the state plaintiffs.

Although Medicaid began in 1965 as a health care entitlement program for the very poor and indigent, it has grown tremendously through the decades in size, scope, and cost. It is jointly funded by the federal and state governments. Expanding Medicaid was initially quite attractive to states because federal taxpayers pick up at least half of a state program's costs and in most cases much more.[30] However, increased Medicaid spending in recent years, particularly during the recent recession, has crowded out other state budgetary priorities such as education and general government. State budgets remain plagued by the combination of flat or declining revenue, unfunded pension liabilities, and unmanageable deficits.

ObamaCare severely compounds those problems. It requires states to expand Medicaid to cover anyone under the age of sixty-five with an annual income at or below 133 percent of the Federal Poverty Level (FPL) by 2014.[31] The new health care law also imposes a "maintenance of effort" requirement that prevents states from changing any of their standards for who is eligible for Medicaid, based on the date the PPACA was signed into law. This requirement will operate until new state-based health benefits exchanges offer coverage to at least some of those populations (among others) beginning in 2014 for adults and until October 2019 in the case of children already eligible for a state's Children's Health Insurance Program or Medicaid program.

Initially, the expanded Medicaid coverage for "newly eligible" state residents that begins in 2014 would be paid entirely by federal taxpayers (many of whom, remarkably enough, are also state taxpayers!) from 2014 to 2016 and then phased down to 90 percent federal and 10 percent state financing by 2020. However, a number of states have complained vigorously that their current budget problems mean that they cannot afford *any increase* in their current Medicaid costs and actually need more flexibility to *reduce* them. Instead, ObamaCare locks in their state budgets and holds them hostage to these new federal mandates. The flexibility of states to trim or eliminate some "optional" Medicaid benefits is limited, and the only other budget-saving approaches available to them are to reduce payment levels to doctors and other Medicaid providers. State officials are also concerned that federal officials will interpret the size of "newly eligible" Medicaid beneficiaries narrowly and thus pass more budget costs on to state taxpayers. They also worry that many millions of people who already are eligible for Medicaid, but who aren't signed up, will enroll in state Medicaid programs in the next few years as they see the individual mandate—and its financial penalties—looming.[32]

COVERAGE ON THE "CHEAP"

Undoubtedly, the main reason Congress decided to require states to enroll millions more people in Medicaid is that it made the PPACA's massive insurance coverage expansions appear to be less costly on the federal ledger. Medicaid traditionally claims to insure many lower-income, nondisabled Americans under the age of sixty-five at a much lower cost per person than does private health insurance—largely because it pays the health care providers treating Medicaid beneficiaries as little as half as much! Of course, this also leads to much lower levels of participation in the Medicaid program by private doctors and produces serious problems of access to quality health care. However, congressional leaders and the Obama administration preferred to cover as many people as possible by the PPACA at the lowest budgetary cost in order to secure support for passage of the legislation. They wanted to pretend that the law would cost no more than $1 trillion. As a result, about half of those newly insured will come through Medicaid mandates and subsidies. These appear to stretch federal dollars to cover more people[33]—but millions more people will be confined to the Medicaid ghetto.

WHEN CHANGING THE RULES BECOMES UNCONSTITUTIONAL COERCION

In this context, hard-pressed state officials (particularly those representing the twenty-six states in the Florida lawsuit) claimed that the federal government imposed new and onerous burdens on their Medicaid programs in the PPACA. They said those burdens were the equivalent of unconstitutional commandeering of their own sovereign government processes and decision making. Their legal argument is as follows.

The federal government can traditionally "bribe" the states by offering federal spending assistance if the states will participate in federal programs. Washington may impose conditions on how states implement the federal programs in which they have voluntarily agreed to participate, *after* those conditions are stated clearly and unambiguously. However, the Supreme Court noted in *South Dakota v. Dole* (1987) that "in some circumstances the financial inducement offered by Congress might be so coercive as to pass the point at which 'pressure turns into compulsion.' "[34] In that case, such commandeering would violate the autonomy of states guaranteed in the structure of the Constitution.

Medicaid has been an ongoing relationship between the states and the federal government over a number of decades. Any "new" conditions in the PPACA that substantially modify the states' role should be compared to the conditions in place when the initial grant of federal funding secured state participation. Could states have reasonably foreseen those future changes? Although it was relatively easy and attractive for states to join the Medicaid program, they increasingly became "locked in"—even as their share of program costs grew too expensive under the new federal conditions—because leaving the program would forfeit more federal matching dollars than most individual states contributed. As Vanderbilt law professor James Blumstein explained recently, "It makes it excruciatingly difficult politically to phase down a program like this because on the way up, one dollar may get you as many as three dollars; on the way down, you save one state dollar but you lose three program dollars."[35]

The PPACA essentially dictates an all-or-nothing feature in the "choice" of continued Medicaid participation by states. Everyone under 133 percent of FPL must be included by 2014. The states have little, if any, role in determining the richness of the benefits and eligibility standards for their present and future role in Medicaid, even as it occupies a larger and larger share of their own budgets and moves well beyond its roots as a poverty assistance program. All they do is serve as "cash cows" and local branch offices of an expanded federal program that is designed in, and controlled from, Washington.

Did Congress go too far in substantially modifying on its own the original terms and expectations of the relationship between the federal and state governments regarding Medicaid?[36]

Judge Vinson was not convinced by the argument that Medicaid participation by states is no longer "voluntary" and that its new conditions might constitute improper commandeering rather than just "a hard political choice."[37] Although he acknowledged that there is a line somewhere between mere pressure and impermissible coercion of state governments, his final ruling determined that current case law failed to support the states' coercion claims. Other potential lawsuits by one or more other states not involved in the Florida case may raise this issue in the future.

ARE EXCHANGES ANOTHER
FORM OF COMMANDEERING?

The other element of anti-commandeering and Tenth Amendment challenges to the PPACA involves how it tries to impose federal standards for health benefit exchanges on the states and ensure their participation in administering them. In *Florida v. HHS*, the plaintiffs argued that the act forces them to operate a health benefit exchange in their state "under threat of removing or significantly curtailing their long-held authority." However, Judge

Vinson considered the exchanges under the law to be "voluntary" and not to compel states to regulate the private conduct of their citizens. He dismissed that portion of the states' complaint and ruled that the exchanges were structured as the type of cooperative federalism authorized under the Constitution.[38]

However, the state-based challenges to the constitutionality of the exchange structure and its requirements deserve further examination. The idea of helping consumers shop for health insurance through exchanges has broad support, but the rules for the exchanges could vary dramatically. The new health care law requires that each state "shall" establish an American Health Benefit Exchange no later than January 1, 2014, to facilitate the purchase of a "qualified" health care plan for individuals and small businesses.[39] States have some modest degree of flexibility in *how* they will comply with the new federal requirements, but the exchanges remain a key mechanism of centralized regulatory control over health insurance. They are essential to its intricate web of price controls and large-scale income redistribution by way of taxpayer subsidies. The new health care law gives broad power to the HHS secretary to supervise and control how states establish their exchanges. A long list of federal requirements for exchanges limits which plans can qualify and dictates the levels of benefits they must offer.

CHOICE, OR NO CHOICE AT ALL?

The exchange section of the PPACA was drafted in an unusual manner that makes it potentially vulnerable to a different type of constitutional challenge. Although Section 1311 establishes the command-and-control instructions for state-run exchanges, it comes far too close to explicitly improper commandeering of state officials to survive on its own. Its artificial distinction in having a state "choose" between establishing an exchange as a department of its own government and doing so as a not-for-profit entity created by the state makes little real difference. Whether the exchange is created by the state or contracted out, it will be following and implementing federal law to one degree or another. As a constitutional backstop, the PPACA also establishes in Section 1321 "alternative" exchanges for a state that would be run by federal officials if the HHS secretary does not certify that a state has established its own qualified exchange (meeting federal requirements) by January 2013.

Health benefits attorney Thomas Christina explains that states are really being told that either they can adopt the substance of HHS regulations for an exchange as their own state statute, or they can adopt them in some other fashion that pleases the HHS secretary. That "choice" isn't really a choice because HHS will establish an exchange in any state that does not directly capitulate. Given the additional political lubricant of "free" federal taxpayer subsidies for state officials who do as they are instructed, this thinly disguised federal affront to the principle of noninterference in state self-governance and political accountability appears worthy of further serious constitutional challenge.[40]

STATES FIGHT BACK BEYOND THE COURTS

A number of states are finding other ways to resist ObamaCare. For example, voters in Arizona and Oklahoma approved amendments to their state constitutions in November 2010, modeled on the Freedom of Choice in Health Care Act developed by the American Legislative Exchange Council. They voted to preserve the freedom of their residents to provide for their own health care. The constitutional amendments would prohibit any law or rule to compel anyone in their state to participate in any particular health care system. They would also ensure the right of a person or employer to pay directly for lawful health care services without being required to pay penalties or fines to do so. A similar, but somewhat more far-reaching, proposed constitutional amendment was narrowly defeated by Colorado voters last November. Other states have simply passed state laws rather than constitutional amendments, almost all expressing disapproval of ObamaCare's individual mandate. Several states have also refused to accept federal funds to start plans to set up state-based health benefit exchanges or are considering developing alternative market-based mechanisms of their own.

More ambitious responses remain under consideration in a number of other states, including the creation of interstate compacts between a group of states and the Congress as federal law that would create an alternative state-based regulation of health, including an option to drop out of the ObamaCare federal regime entirely.[41]

RESTORING THE CONSTITUTIONAL BALANCE

In 2011, it remains unclear whether we will be able to retain the constitutional rights that are threatened by ObamaCare. The rulings of lower federal courts have been divided, and the ultimate legal decision rests with the U.S. Supreme Court to determine, probably in 2012. The initial round of court cases has, at a minimum, reacquainted Washington officeholders and their appointed regulators with the original structure of the Constitution, which was designed to protect and preserve our fundamental rights and freedom. The Constitution relies on the essential role of the states in enforcing the balance between the national and local spheres of self-governance that helps ensure individual liberty.

Most of the last century and the beginning of the current one witnessed a steady expansion of the power of the federal government over individual Americans and its intrusion into traditional areas of state governance. ObamaCare aims to continue this erosion of long-standing constitutional limits on the reach of federal power. It also hopes to enlist state governments as political buffers that will absorb some of the political shock waves and crushing financial burdens of a complete federal takeover of our health care. Full implementation of this combination—vastly expanded Medicaid coverage through carefully controlled state government instruments, much tighter national regulation of "private" health insurance in heavily subsidized "exchanges," and the legal requirement that everyone (or their employers) must purchase federally defined products (or their equivalents)—would amount to a "public plan" for universal coverage by means that are only partly disguised and slightly less direct.

DIMINISHING POLITICAL ACCOUNTABILITY

Justice Antonin Scalia's opinion in *Printz v. United States* (1997), which struck down federal legislation that required states to conduct background checks on prospective purchasers of guns (for lacking constitutional authority under the Necessary and Proper Clause), warned us of the dangers of congressional laws that require states directly to enforce federal laws and diminish political accountability:

> By forcing state governments to absorb the financial burden of implementing a federal regulatory program, Members of Con-

gress can take credit for "solving" problems without having to ask their constituents to pay for the solutions with higher taxes. And even when the States are not forced to absorb the costs of implementing a federal program, they are still put in the position of taking the blame for its burdensomeness and for its defects.[42]

In a similar manner, both the mandate on individuals and the related mandate on employers to pay for "private" coverage under the new health care law essentially treat health insurance as "a revenue-raising mechanism" that the federal government can use to pay for health care (and redistribute its costs). This raises the constitutional question as to whether private health insurance companies can be conscripted to act as branch offices of both the IRS and HHS—first to collect taxes and then to pay benefits—without the standard political controls applied to official government agencies.[43]

A full government takeover bid would fail if brought before voters honestly and openly. But using private parties as both tax collectors and contributors of their own mandated funds might make the explicit budgetary costs appear somewhat lower. The entire scheme makes one wonder what else Congress can make Americans do if it can make us buy health insurance.

There ought to be a law against this—and there is. It's called the Constitution. Let's use it, again.

WHAT WE SHOULD DO INSTEAD TO GET REFORM RIGHT

PUTTING PATIENTS FIRST

Though ObamaCare is law, it is not settled policy. Most Americans continue to say that President Obama and the 111th Congress got health reform wrong—very wrong.[1]

Faced with this growing criticism, supporters insist that they can fix the law with more legislation and more regulation. We disagree. The problems with ObamaCare cannot be fixed because they are woven into its fabric. The law is fundamentally and structurally flawed and cannot be repaired or improved. It must be repealed and replaced.

You simply can't build a patient-centered health care system on ObamaCare's foundation of bureaucracy and central planning.

Throughout this book, we have described in detail where this law will take us—to higher costs, lower quality, long waits for care, and limits on access to new treatments and medicines.

The very real problems with health care in the United States can be solved, but not with ObamaCare. The key is to empower patients, not the government.

We are at a crossroads: either we can move to a truly market-driven health care economy that puts consumers in charge of choices, or we can continue to build one that puts government at the center. ObamaCare will be governed by hundreds of new boards and commissions, thousands of regulators, and tens of thousands of pages of regulation.

Instead, we support a system that builds on the core strength of our market economy and puts consumers in charge. You would have new powers and the tools to demand better choices in a competitive marketplace that offers more affordable health insurance and policies that suit your needs.

But to get there, Congress must fix some of the fundamental flaws that are keeping you from being able to own and control your own health insurance and therefore make more of your own health choices. You could pick policies that would allow you to keep young adult children on your policy, cover preventive care with no out-of-pocket costs, and have no limits on what your policy will pay. The policies would likely be more expensive; the difference is, you would decide whether they are worth the cost.

Our current system has built so many barriers to allowing you to make these choices that many people think it just can't be done. "It's better to have ObamaCare than the system we have now," some say. But the future under ObamaCare is bleak. Instead, we can build a bright twenty-first-century health care economy. The key is to give power to the people!

FACTS OF LIFE

Between now and 2013, several facts of life remain. The current health overhaul law is on the books, and bureaucrats in the Obama administration are grinding away to put the regulatory machinery into place. It's unlikely that Congress could muster the votes in the 112th Congress to overturn President Obama's predictable veto of a bill that would overturn the law. And it will be hard to convince independent voters that a sweeping repeal of ObamaCare is a good idea until they see better ideas being offered. Free-market reformers must offer positive solutions, not a status quo that is unsatisfactory to many Americans.

Ironically, supporters of ObamaCare argue that their plan builds on the free market and consumer choice and that it absolutely, positively is not a government takeover of health care.

They did use some of the names of sensible concepts and ideas that free-market analysts have been advocating for years. For example, they agreed that, in our increasingly mobile economy, Americans should have portable insurance that they can take with them as they move from job to job. To do that, subsidies for health insurance premiums are portable. They also agreed that you should be able to pick your own health insurance, just as you pick other types of insurance.

The architects of ObamaCare gave a rhetorical nod to these ideas by providing tax credits to help people afford premiums and by creating new marketplaces—or "exchanges"—where people can shop for portable insurance policies.

But ObamaCare took these market-based ideas and twisted them beyond recognition into a bureaucratic knot. The exchanges are weighed down with rules, regulations, and government restrictions and crushed under a mountain of bureaucracy. The only things that remain of free-market ideas are the labels: tax credits, portability, and consumer choice. We need to peel off those labels and put them on the right policies. We need to get the incentives right with policies that reward people for making the right choices and provide a strong safety net to help the most vulnerable people.

In a properly functioning market, millions of people make billions of choices every day about what they do, where they go, and how they spend their money. Everyone offering a service or a product in the market gets these signals, and they respond by providing consumers with more of what they want and less of what they don't. Different companies offering the same services compete with one another to do a better job so you will pick them over their competitors. Why on earth would we want a system, especially in something as personal as health care, where all of those market signals are lost and doctors have to respond to regulators, not you?

PUTTING CONSUMERS IN CHARGE

The central pillar of free-market reform is a vigorously competitive health sector.[2] Many of the problems the country is facing involving health care costs could be addressed by encouraging much more competition and giving consumers greater control over resources and decisions involving their care and coverage.[3] Greater transparency in both prices and performance, along with a larger choice of options, would drive out insurers that charge more for their products than they are worth to consumers. This delivers much better results than having government officials throw rhetorical stones at insurers just to score political points.

Unfortunately, the lack of competition in health insurance in many states limits the options for coverage, overregulation drives up costs, and our structure of financing health insurance gives consumers little power to make choices. ObamaCare will make those problems worse.

Though health care is different from other sectors of our economy and requires special consideration, there are many areas in which consumers want to and can have more control over their health care choices—especially regarding the kind of health insurance policy they want and how much they are willing to pay for it.

Competition can work when consumers are truly engaged in getting better value for their health care dollars.[4] Government needs to get out of the way, not build more barriers.

"There will be no meaningful cost control until we are all cost controllers in our own right."[5]

—Indiana Governor Mitch Daniels

MONEY TO THE PEOPLE

Health care costs are rising relentlessly, and this is forcing up the cost of health insurance for individuals, employers, and government. Many people with existing medical conditions don't have access to affordable insurance. And too many families are uninsured and are just one illness or accident away from financial disaster. Reformers must offer new ways for people to get health insurance, ways to keep costs under control, and ways to reform Medicare and Medicaid with real solutions rather than shell games.

This has to start with a new vision, one aligned with a twenty-first-century economy in which you, rather than government bureaucracies, have control over your own resources. Our vision is to build on a process that switches existing health care benefits to "defined-contribution" programs. This concept can be applied to taxpayers' financing of all three major insurance coverage platforms in the United States—Medicare, Medicaid, and private health insurance.

It is a concept already familiar to most workers. The defined-contribution revolution is already well under way in private retirement programs. In 1985, about 80 percent of all full-time U.S. workers were enrolled in company pension plans that offered them a defined benefit. Those plans promised workers that the company would pay them a pension of a certain amount after they retired.

By 2006, the share of those workers in defined benefit plans had dropped to just 20 percent. The last quarter century has seen a revolution in 401(k)s, IRAs, Keogh plans, and other defined-contribution plans that allow retirement funds to be portable and controlled by individual workers. Health insurance could and should offer the same opportunity.

Defined-contribution payments are a different path to comprehensive health reform. Payments would be made more directly to people rather than laundering, hiding, and redirecting spending through third-party middlemen. Having more control over their health care spending would empower and encourage consumers and patients to make better health care choices. It would stimulate more innovative and accountable competition by health care providers. And it would encourage us all to save and invest so that we can afford to pay more for health care when it delivers more value but redirect our resources elsewhere when it delivers less.

This wholesale shift has not yet occurred in health care because federal entitlement and tax policies are in the way. They have propped up arrangements that emphasize government support over patient satisfaction.

Here's How the System Works Now

If you work for a company that provides health insurance, you get a generous tax break if your employer writes the check for it. The premiums for your health insurance are taken off the top, or excluded, from your income before your employer calculates the taxes you owe.[6] No matter how much your employer spends on health insurance premiums, you don't pay any taxes on that part of your compensation. (The insurance is not a gift from your employer; it is part of your pay package. The more it costs, the less you have in take-home pay.)

It's cheaper to buy health insurance with these pre-tax dollars than to buy it on your own with after-tax dollars.[7] The tax savings can be worth several thousand dollars a year. So over the decades, consciously or not, tax policy has provided a big financial incentive for tens of millions of Americans and families to get health insurance at work.

Incentives to economize, among employers as well as workers and consumers, are muted because the spending—and tax breaks—are invisible and without limit. Today, if you earn a salary of $75,000 for your family and your employer also enrolls you in a $13,000 health insurance policy, the tax break for that coverage is worth about $5,200.[8]

But people pay a high price for this generous health insurance at work. Take-home pay has risen modestly over the last decade—just 38 percent on average— in large part because so much of your compensation has been gobbled up by expensive health insurance, which has gone up by 131 percent over the same period.[9] Your employer isn't paying for the bulk of your policy; you are. Your employer has been taking the money out of your overall compensation package to pay your health insurance premiums. At some point, we just have to say ENOUGH!

The same thing is happening with Medicare and Medicaid. If you are in one of those programs, you have little or no control over the price of the health care you are consuming since your doctors and hospitals send the bills to government (or your supplemental insurance plan). The new health care law would make matters worse by moving millions more people into heavily subsidized, taxpayer-supported programs (through either state health benefits exchanges or Medicaid). These are the very kind of open-ended financing arrangements that are at the heart of today's cost problem.

We can't go on like this forever. At some point, someone has to put on the brakes. ObamaCare does it with price controls, which will ultimately turn into rationing. We want to do it by engaging the power of the free market so that health care is better, more convenient, *and* less expensive.

Here's How It Could Work

Every American household—whether working or not—would be eligible for a health care credit.[10] The credit could be used only to purchase health insurance and health care services. Any household that didn't buy coverage would lose the entire value of the credit. The number choosing to do so would likely be very small. The credit would be available to you whether you get your health insurance on your own, at work, or through other kinds of groups. The value of the credit would be determined by Congress, but, depending upon its size, could be largely financed by modernizing the tax treatment of health insurance. Moving toward a universal, fixed-dollar credit would be a different, but real, alternative to the new health law and could actually be a path to health insurance for everyone.[11] The credit also would begin to break down the obsolete regulations we have now that tie health insurance to your job. You would be in charge of choices, and your health insurance could stay with you.

This Works for Medicare and Medicaid, Too

Something similar could and should be done in Medicare. In fact, it's already been done in a part of Medicare: the prescription drug benefit, which was added to the program in 2003. Medicare beneficiaries decide for themselves what kind of drug insurance they want to purchase each year. The government's contribution is based on the average price charged by participating drug plans in a region, and it remains the same regardless of which plan the beneficiary selects. If seniors select plans that are more expensive than the average price of all plans offering bids each year to provide drug benefits, they must pay the additional premium out of their own pockets. Many select less expensive plans available to them because scores of insurers compete aggressively with one another for their business every year. The result is that federal spending on the program is some 30 to 40 percent below initial projections.[12]

As new retirees turn sixty-five and enter Medicare, they should be given the same control they now have over drug coverage for their entire Medicare benefit. The resulting competition among private plans and the traditional Medicare fee-for-service program, as well as among those providing health care services to the beneficiaries, would work far better than top-down government regulation alone to hold down costs and premiums. It would also make insurers much more attentive to the needs of patients, as seniors who were dissatisfied with how they were treated could take their business elsewhere at the next available opportunity. (Open enrollment seasons would take place at least once a year.)

Medicaid—the health insurance program for low-income

Americans—locks participants into a system of insurance that is largely separate from what everyone else experiences. Medicaid's payments to doctors and hospitals are so low that many physicians simply refuse to see patients with Medicaid coverage. Medicaid is also not coordinated with employer insurance, so when low-income people get higher-paying jobs, they often lose Medicaid without the guarantee of getting coverage at their job.

All this could be fixed by giving the nondisabled who are enrolled in Medicaid the same health credit given to all other working-age Americans. That would serve as their base amount of taxpayer assistance to help them get private coverage. What is now spent on them in Medicaid could then be used to provide additional help with premium expenses as well as out-of-pocket costs when they need health care. This approach would allow Medicaid participants to get the same kind of insurance—and thus access to the same doctors and hospitals—as everyone else. It would also mean that they could keep the same insurance even when they move up to higher-paying jobs with new employers.

Building an effective marketplace for health care requires some additional steps beyond reforming how employers, Medicare, and Medicaid pay for insurance. We also need more transparent information about prices and quality, better state regulation of insurance markets, and further assistance for beneficiaries making choices about what will be best for them.[13] Nevertheless, it's clear that the key reform that is needed is one that puts you—and your fellow Americans—in charge of the money. That's the way to slow rising costs while also improving, not compromising, quality.

THE RYAN ROADMAP

It is impossible to discuss comprehensive health care and entitlement reform without mentioning the most comprehensive proposal to accomplish what's needed: the "Roadmap" offered by Representative Paul Ryan,[14] the new chairman of the House Budget Committee.

In drafting his Roadmap, Congressman Ryan's explicit objective was to restore long-term balance to the government's budget without resorting to huge tax increases by building a network of affordable entitlement programs that will not crush America's entrepreneurial initiative.

The Roadmap puts you in charge of your health care by simultaneously reforming Medicare and Medicaid, as well as the tax preference for employer-paid insurance. His plan lays the foundation for restoration of long-term balance to the federal budget.[15]

In the Roadmap, Americans age fifty-five and younger would enter into a restructured Medicare program that paid fixed dollar amounts to the cost of their health insurance. The tax exclusion for employer-provided health insurance premiums would be repealed and converted into fixed-dollar refundable tax credits for all households with a member under the age of sixty-five.[16] States would choose either to allow their Medicaid recipients (except for those who are disabled or receiving long-term care assistance) to participate in the tax credit plan (augmented by additional low-income assistance) or to use federal block grant funding to restructure their Medicaid programs with greater flexibility.

Over the long run, the massive run-up in debt that would

occur under the current law would be avoided entirely, even as taxes would not be increased.

But even those numerical indicators do not fully highlight the importance of what Representative Ryan has proposed. The Ryan plan would convert tens of millions of Americans into cost-conscious consumers. The combined effect of doing so across age groups and incomes would create tremendous competitive pressure on insurers and health care providers to deliver much better value at less cost.

The Ryan Roadmap is thus not just a budget plan. It is also a plan to transform American health care. It is built on the fundamental principle that with you and your fellow Americans in charge of health care dollars, we can get much better health care at lower cost than we are getting today.

ENSURING THAT SICK AMERICANS AREN'T LEFT OUT

Even if we fix U.S. health care by putting consumers in charge, there will still be a problem with making sure those who are especially sick can get good insurance too. Today, there are an estimated two to five million Americans who have a serious "pre-existing condition" that would deny or limit their access to health insurance coverage.[17] Private insurers in the individual market must take the much higher risk—and costs—of those ailments into account when setting premiums, or limiting initial coverage for them.

In fact, the president is holding up the plight of these Americans as the main rationale for the government takeover built into ObamaCare. While describing the plight of a young woman in the audience of a rally he attended in April 2010, President Obama told the crowd, "If [opponents of the law] want to look at Lauren

Gallagher in the eye and tell her they plan to take away her father's ability to get health insurance . . . they can run on that platform."

The choice the president presents is a false one. And his solution isn't working. We have a problem that needs to be addressed, but turning over the entire system to government control is not the answer.

Ironically, ObamaCare included an ill-conceived version of what is needed to actually ensure that sick Americans can get affordable insurance. ObamaCare created a $5 billion "high-risk pool" program that is poorly designed, hastily constructed, and severely underfunded. About 375,000 were expected to apply by the end of 2010, but only about 8,000 people had enrolled by then.[18] Clearly, this program needs to be rethought.

The lesson to be learned from ObamaCare's early experience is not that high-risk pools don't work; it's that they won't work if they are built to fail. For starters, they have to be reliable and well funded to get people to sign up. They should provide assistance to people who find themselves, through no fault of their own, looking for insurance as individuals even though they have a history of high medical costs. The high-risk pool should help them pay their premiums, and state regulations would place an upper limit on how much they can be charged directly. The pools need to be coupled with stronger protections for those maintaining continuous insurance coverage so that they don't find themselves needing high-risk assistance at all.

High-risk pools can address preexisting conditions without the costs and burdens of the heavy-handed federal regulation of insurance planned for 2014.[19] In short, we can do more by doing less, and the solution can be transparent, well targeted, and adequately funded.

High-risk pools that deliver what they promise will naturally be more expensive. But compared to the sweeping burdens of

ObamaCare, they will cost much less and do less damage to the rest of the private health care market, which many Americans prefer and from which they still benefit greatly. High-risk pools can be the foundation of what it means to replace, not just repeal, ObamaCare's flawed prescription for health care policy change.

ADDITIONAL IMPORTANT REFORMS
AT THE STATE LEVEL

Putting consumers in charge and protecting those with preexisting conditions are the crucial steps needed to broaden coverage and slow the pace of rising costs.

But some other steps are needed as well. In our federalist system of government, states should be given the latitude to take the lead in implementing such changes in ways that are most acceptable to their citizens.

For instance, consumers should be given new opportunities to find and purchase the coverage that best fits their needs. In that regard, states should be given greater flexibility to put into place insurance marketplaces that would give consumers accurate and comprehensive information about their options, as well as simplify the process of enrollment. Those marketplaces should emphasize more consumer assistance, not tighter insurance regulation. The Utah Health Exchange is a model. Exchanges can be a mechanism to aggregate defined contributions and allow consumers to purchase health insurance with pretax dollars.

The federal government should also allow the states to use those mechanisms to fundamentally transform their Medicaid programs into consumer-choice models.

The states should also open the sale and purchase of health

insurance to consumers in other states. This would help residents in states with costly and excessive regulations to find affordable coverage elsewhere and would drive down costs.

States also must reform their medical malpractice laws to remove the distortions associated with arbitrary and unlimited jury awards. Today, lawyers shoot for the sky and hope they come up winners. The limited number of cases where they do hit the jackpot is causing doctors to do whatever is necessary to avoid the same fate. The result is billions and billions of dollars' worth of unnecessary medical treatments. Sensible tort reform would allow doctors to practice sound medicine and still protect patients from genuine malpractice.

THE PATH FORWARD

Moving forward, leaders in Congress must start to work on reform that is compatible with our nation's political culture, the values embodied in our Constitution, and our free-market economy.

The challenges are enormous, even without the upheaval of ObamaCare. Millions of baby boomers will soon be signing up for Medicare, putting new pressures on the system. The costs of government entitlement programs threaten to squeeze out other public services provided by federal and state governments. Millions of people continue to lose their health insurance when they lose or change jobs. And the cost of health care and insurance coverage continues to rise.

It is not possible for a government bureaucracy to know how to solve all those problems or to address the diverse needs of 300 million Americans. What's needed is reform that puts you and your fellow citizens, not government, in charge of the health care dollars. That's the way to ensure your needs are addressed, with better health care at less expense for all of us.

A CHECKLIST FOR A STEP-BY-STEP APPROACH
TO HEALTH CARE REFORM

Private Insurance

- Offer people a health credit to purchase coverage, either on their own, through an employer, or through other groups.

- Allow greater flexibility in health benefits: consumers, not regulators, should decide what their health plans cover and not be forced into one-size-fits-all, government-determined standard plans.

- Provide portability of health insurance and greater competition by allowing cross-state purchase of health insurance.

- Allow states to develop market mechanisms to help consumers find and enroll in the insurance that best meets their needs.

- Ensure more secure renewal of health insurance that is guaranteed so people who have insurance can keep it, and others without it will be encouraged to get insurance and maintain their coverage continuously.

- Provide greater financial assistance to the states to create more functional high-risk pools or state risk-transfer pools that allow people with preexisting conditions to purchase more affordable health insurance.

- Reform the medical malpractice litigation process at the state level.

Public Programs

- Put the savings from Medicare reform into saving Medicare.

- Convert Medicare for new enrollees into a market-based, consumer-choice program in which the beneficiaries select the coverage that best suits their needs with fixed support from the government.

- Allow people to escape from Medicaid by giving them health credits that they could use to purchase private coverage.

- Provide more flexibility to the states in running Medicaid programs so they can get the best value for taxpayers' dollars, including allowing Medicaid beneficiaries to enroll in state-designed consumer-choice models.

- Provide more options for Medicaid recipients, Medicare beneficiaries, and others on public programs to escape the restrictions that inevitably come from price controls and government micromanagement.

WHAT YOU CAN DO TO PUT THE BRAKES ON OBAMACARE

POPULAR PRESSURE CAN MAKE A DIFFERENCE. YUU HAVE POWER!

Dozens of members of Congress were elected in 2010 on a pledge to repeal ObamaCare. They will be looking for every opportunity to show progress toward that goal before they face the voters again in 2012. After voting for repeal, they now are working to defund, dismantle, and delay whatever they can in the new law, as well as to shine a light on the many regulations the Obama administration has been issuing to implement it. They also are working to give states more power to defend their citizens against ObamaCare's sweeping mandates.

The states will be the power centers in 2011 and 2012. Many governors who were elected in staunch opposition to the federal health overhaul law are trying to build firewalls against Obama-Care's costs, mandates, and intrusions. House and Senate leaders plan to work closely with governors and state legislators to try to give them more flexibility and authority to do their own reform. Governors were among the first witnesses at House committee hearings in early 2011 to explain the impact of ObamaCare on their states and their recommendations for better solutions.

What can you do? The most important thing is to make

sure your voice is heard. What President Obama and his friends in Congress want most is for you to be silent and pretend that ObamaCare isn't so bad after all. Whether you support or oppose the new health law, or like some parts and have problems with others, it's time we all had the vigorous public debate that was supposed to happen but that was bypassed in 2009 and 2010. We must keep up the fight! Here are fifteen things you can do to make your voice heard:

1. **Stay informed**. Arm yourself with information so you can help your fellow citizens understand what this law will mean for them and the future of our country. See below for a list of key facts you will need.

2. **Recommend this book** to your friends and neighbors so they too can learn what is really in the law and the impact it will have. And visit our website at www.WrongForAmerica Book.com for continuing updates.

3. **Write letters to members of Congress** to let them know how ObamaCare will affect you, your family, and your business. Tell them what's happening with your health care costs and let them know any way in which your coverage is changing—such as coverage for dependents being dropped because of the costs of the new mandates in ObamaCare.

4. **Write to the president** to let him know your views on health reform and how it is being implemented.

5. If you learn of a new federal regulation that impacts you, **send in a comment**. The administration is legally required to consider your views.

6. **Attend town hall meetings** in your district when your federal and state representatives organize them to get informed, and make your voice heard.

7. If you are on Medicare or Medicaid and are having trouble finding a doctor, **let your elected representatives know**.

8. When there are votes in Congress on repealing or preserving the law or any of its parts, **call and write your representatives** to tell them what you think and how you want them to vote.

9. **Contact your state's governor and senators and your representative** to let them know your views as they face the challenges and expense of implementing many of ObamaCare's provisions.

10. **Call in to talk-radio programs** with specific examples of and comments about how reform affects you.

11. **Write letters to the editor** of your local paper, respond to reporters who offer their e-mail addresses, and post comments on blogs.

12. **Start a Facebook page** to give your friends a chance to write their own stories and tell them to others.

13. **Sign up for free reports on health reform**. We four authors all have websites you can visit and regular reports we can send you to stay informed, and we can recommend others. (See our list at the end of this chapter.)

14. **Set up or join a discussion group** with friends and associates. You can exchange information and invite experts to come talk with you.

15. **Attend meetings and join demonstrations**. You are still free to assemble.

And if you can do it, consider running for office. The huge 2010 electoral wave swept nearly a thousand new members into state and federal offices, and 2012 is shaping up to be an even more consequential election.

THE FACTS YOU NEED

To help in this fight, you need to arm yourself with the facts.

More than half a trillion dollars in new taxes will be passed on to consumers in higher health care costs and insurance premiums. Business owners are aghast at the avalanche of mandates and new costs, which are stifling job creation. Seniors know you can't take $575 billion out of Medicare and make their coverage more secure. And working-age citizens, particularly young people, are becoming concerned about the cost of complying with the mandate that everyone must have government-approved health coverage.

Here are ten facts that you need to know about ObamaCare, summarizing a few key points from the book:

1. **Higher costs**: If you buy a family policy in the individual market, you can expect to pay $2,100 more a year in premiums by 2016 than you would if Congress had not passed ObamaCare. That means the policy will cost on average $15,200 in 2016 for health insurance with the law and $13,100 without it.[1] The federal government's chief Medicare actuary, Richard S. Foster, recently reported that Americans will spend an average of $265 more per person a year on health insurance as a result of the health care law by 2019.[2] That works out to more than $1,000 more a year for a family of four than if the law hadn't passed.

2. **Tax hikes**: ObamaCare includes $569 billion in tax hikes to fund its expensive subsidies. Medicare actuary Rick Foster finds that the hundreds of billions of dollars in new fees and excise taxes will "generally be passed through to health consumers in the form of higher drug and devices prices and higher insurance premiums."[3]

3. **Seniors losing coverage**: Mr. Foster estimates that more than 7 million seniors with Medicare Advantage coverage will lose their current coverage because ObamaCare cuts reimbursements to that plan by $145 billion.[4] Mr. Foster confirmed that the health care overhaul law will result in "less generous benefits packages" for seniors on the popular Medicare Advantage program and that the coverage will cost them more. He estimates that seniors' costs will go up by $346 a year in 2011 and by as much as $923 by 2017.[5]

4. **More bureaucracy**: ObamaCare establishes an estimated 159 new boards, advisory commissions, and programs, all of which will weave new webs of bureaucratic red tape. The Congressional Research Service says the exact number is "unknowable."

5. **Higher government spending**: Government health care spending will grow faster than without "reform." Mr. Foster estimates that the health care law will increase overall national health expenditures by $311 billion between 2010 and 2019.

6. **Larger deficits**: ObamaCare starts collecting new and higher taxes in 2010, but the entitlement spending doesn't start until 2014—six years of spending and ten years of taxes. The real cost of ObamaCare will be at least $2.3 trillion when a full ten years of spending is counted. Former CBO director Douglas Holtz-Eakin and our coauthor Jim Capretta write that employers will have strong incentives to move as many as 35 million workers out of employer plans and into subsidized plans. They estimate that this would add about $1 trillion more to the total cost of ObamaCare over the next ten years.[6]

7. **23 million uninsured**: The bill will leave an estimated 23 million people without insurance by 2019, not even close

to the promised goal of universal coverage.[7] This will be a serious problem for hospitals, which still will be treating uninsured people, including illegal immigrants, but will be facing federal cuts in payments for uncompensated care.

8. **Insurance death spiral**: Individuals will be required to purchase health insurance, and younger workers will be forced to pay higher premiums to subsidize older Americans. If younger and healthier people choose to pay penalties rather than buy expensive insurance, premiums could soar for those left in the insurance pool.

9. **Losing your coverage**: As many as 80 million to 100 million people could find themselves with different coverage by the time ObamaCare fully takes effect, despite the president's repeated promise that "you will be able to keep your health care plan."[8]

10. **Medicare provider losses**: In a report in August 2010, Medicare's Office of the Actuary said that the number of hospitals, nursing homes, and hospice centers facing financial losses under the new law would jump to "roughly" 25 percent in 2030 and 40 percent by 2050,[9] significantly impacting access to care for Medicare beneficiaries.

UNDERSTANDING
THE BATTLEFIELD

The court challenges to ObamaCare could be crucial in the overall struggle. It would be an unprecedented abuse of federal power to require citizens to buy government-approved health insurance. The core of the law says that if you don't spend your own money on something the government has deemed desirable, you will be penalized. If that's possible, there's no end to what the federal government can require us to do.

But we can't assume that the courts are going to fix this problem for the country by overturning the entire law. That may never happen. As the court cases proceed, therefore, it's crucial to engage the battle vigorously on other fronts—political, legislative, and regulatory—so that policy makers feel the pressure to change course even if the law withstands court challenges.

That's where you come in. More often than not, lawmakers respond to the signals sent by their constituents. They need to hear from you.

WHAT TO EXPECT IN THE COMING MONTHS

Here is an overview of routes members of Congress can be expected to pursue:

- **Legislative actions** began with the full repeal vote in the House of Representatives on January 19, 2010, and will continue with targeted votes to defund, delay, dismantle, and do direct oversight and investigation, especially into the avalanche of regulations being issued every day. There will be many hearings, floor votes, and numerous legislative offerings over the next two years to help in this educational effort. But members know they can't stop there; they must couple any repeal efforts with a positive, step-by-step agenda for real reform.
- **The political front** will involve continuing educational efforts by outside groups and political candidates to help the American people understand more about the details of

ObamaCare and its impact on the health care sector and
the economy, especially jobs and health care costs. Obama-
Care will inevitably be a major topic in the presidential
campaign.

- **Legal challenges** will likely grow. A *Wall Street Journal*
  editorial[10] encouraged newly elected governors and attor-
  neys general to join the Florida lawsuit, saying "The voters
  showed their loathing for the law on November 2, and a
  large, united legal front of states would increase the chances
  that the courts find it unconstitutional." Six states joined
  the 20 suing the federal government in Florida, and several
  other states, including Virginia and Oklahoma, have filed
  independent court challenges of their own. Expect other
  lawsuits to be brought against specific provisions of the law
  as well.

- **Regulatory roadblocks:** While the wheels of the bureaucracy are rapidly churning out regulations to implement
  the health care overhaul law, legislators have a responsibility to protect the American people to make sure that these
  regulations and the implementation processes are doing
  no harm. Expect Congress to take a number of actions by
  conducting careful review of the avalanche of regulations
  being issued, demanding information from key government officials about their plans, strategies, and spending to
  implement the law.

Finally, watch closely what's happening in the states, as
they will be an active front in the battle against ObamaCare. They
are required by the law to spend significant time and resources
setting up the machinery to implement it. Many states are waiting to see what happens in the courts, others are delaying action

as long as possible, while others are aggressively moving forward to implement the law. Some states are developing parallel plans for the right kind of health care reform. They all are watching carefully what the costs of the dramatic coverage expansion will be and are considering their options.

RESOURCES

All of our organizations provide a wealth of information about the latest developments in the health care field. Please sign up for regular updates and visit our websites often:

American Enterprise Institute
www.aei.org

Ethics and Public Policy Center
www.eppc.org

Galen Institute
www.galen.org

The Heritage Foundation
www.heritage.org

For articles, videos, papers, and commentary on health reform from a number of experts and opinion leaders of interest to free-market advocates, please also visit our other websites:

ObamaCareWatch
www.ObamaCareWatch.org

Health Reform Hub
www.HealthReformHub.org

Fix Health Care Policy
www.fixhealthcarepolicy.com

Continue to follow the conversation about *Why ObamaCare Is Wrong for America* at our website, www.WrongForAmericaBook.com.

Acknowledgments

The authors are grateful to. . .

Our extraordinary editor, Adam Bellow, and the amazingly talented team at HarperCollins, especially executive managing editor Cindy Achar and associate editor Kate Whitenight, for getting the book to press in record time, and to our agent Alex Hoyt for helping us find this great publisher.

. . . The terrific team at the Galen Institute, especially our invaluable research director, Tara Persico, who was tireless in tracking down information, and proofing, editing, and integrating changes from all four authors, and to Jena Persico and Sterling Emerling for their attention to detail in proofing the manuscript and to our whole team for their continuing hard work, creativity, and dedication to our work at Galen. To Jason Fodeman, M.D., for helping explain the impact of ObamaCare on doctors and patients and to our great board of directors and many supporters for their unfailing belief in the cause of freedom.

. . . To Genevieve Wood, vice president for Leadership for America Operations at The Heritage Foundation, for her work in helping to alert people across the country to our book and the real truth about ObamaCare, and to our colleagues at the American Enterprise Institute and the Ethics and Public Policy Center for their support and assistance.

. . . To our colleagues in the Health Policy Consensus Group, including cofounder John Hoff, and to other colleagues from market-based think tanks and universities across the country for their dedication to free-market ideas.

. . . To all of our families who gave up many evenings, weekends, and holidays with us while we wrote and edited the manuscript on tight deadlines, especially our spouses Douglas Turner, Claire Capretta, Mary Miller, and Barbara Moffit.

. . . And to the millions of people who spoke and organized and marched and wrote letters and attended town hall meetings and who did everything they could to make their voices heard during the historic health reform debate of 2009 and 2010. This book is dedicated to you.

Notes

Foreword

1. "The Defined Contribution Route to Health Care Reform," James C. Capretta and Thomas P. Miller, December 2010, www.aei.org/paper/100164.

The Real Story of ObamaCare

1. "Pence Opposes Government Takeover of Health Care," March 21, 2010, www.mikepence.house.gov/index.php?option=com_content&task=view&id=3995&Itemid=63.

2. Jeffrey H. Anderson, "Democratic Pollster: The Economy 'Was Not the Decisive Factor in this Election; Health Care Was,'" November 3, 2010, www.weeklystandard.com/blogs/democratic-pollster-economy-was-not-decisive-factor-election-health-care-was_514761.html.

"Ari Fleischer Looks Ahead of 2010 Midterms to What 2012 Could Bring to Congress," YouTube, November 3, 2010, www.youtube.com/watch?v=3l-ZltAkwwQ.

3. Shawn Tully, "Documents Reveal AT&T, Verizon, Others, Thought About Dropping Employer-Sponsored Benefits," *Fortune*, May 6, 2010, http://money.cnn.com/2010/05/05/news/companies/dropping_benefits.fortune/.

4. Douglas Holtz-Eakin and James C. Capretta, "Resetting the 'ObamaCare' Baseline," December 16, 2010, www.politico.com/news/stories/1210/46428.html.

5. Letter from Secretary Kathleen Sebelius to America's Health Insurance Plans, September 9, 2010, www.hhs.gov/news/press/2010pres/09/20100909a.html.

6. Jeffrey H. Anderson, "ObamaCare Ends Construction of Doctor-Owned Hospitals," January 3, 2011, www.weeklystandard.com/blogs/ObamaCare-ends-construction-doctor-owned-hospitals_525950.html.

Sarah Kliff and Jennifer Haberkorn, "GOP: Repeal Vote This Week . . . Walker Paves Way for Wisconsin Repeal Lawsuit . . . Schwarzenegger Names Belshe, Kennedy to Exchange Board . . . Six States to Watch in 2011," January 3, 2011, www.politico.com/politicopulse/0111/politicopulse404.html.

7. U.S. Department of Health and Human Services, "Approved Applications for Waiver of the Annual Limits Requirements of the PHS Act Section 2711 as of December 31, 2010," www.hhs.gov/ociio/regulations/approved_applications_for_waiver.html.

8. Remarks of President Barack Obama—As Prepared for Delivery to the American Medical Association, June 15, 2009, http://blogs.wsj.com/washwire/2009/06/15/obama-if-you-like-your-doctor-you-can-keep-your-doctor/.

9. "New Consumer Protections Starting As Early As This Fall," HealthCare.gov, a federal government Website managed by the U.S. Department of Health & Human Services, www.healthcare.gov/law/provisions/billofright/patient_bill_of_rights.html#NewConsumerProtectionsStarting.

The new Patient's Bill of Rights regulations detail a set of protections that apply to health coverage starting on or after September 23, 2010, six months after the enactment of the Affordable Care Act. They include:

No Pre-Existing Condition Exclusions for Children Under Age 19
No Arbitrary Rescissions of Insurance Coverage
No Lifetime Limits on Coverage
Restricted Annual Dollar Limits on Coverage
Protecting Your Choice of Doctors
Removing Insurance Company Barriers to
    Emergency Department Services

10. Richard S. Foster, "Estimated Financial Effects of the 'Patient Protection and Affordable Care Act,' as Amended," April 22, 2010, www.cms.gov/ActuarialStudies/Downloads/PPACA_2010-04-22.pdf.

11. The major health care reform legislation that passed on March 21, 2010, and was signed on March 23, 2010, is called the Patient Protection and Affordable Care Act, Public Law 111–148. Two days later, both the House and the Senate passed a second bill, the Health Care and Education Reconciliation Act of 2010 (P.L. 111–152) to make amendments to the first bill. The Joint Economic Committee's Republican staff calculates that the two bills combined total 2,801 pages.

12. This estimate is based on an analysis of the Senate version of the legislation by the staff of the Senate Republican Policy Committee.

13. Curtis W. Copeland, "New Entities Created Pursuant to the Patient Protection and Affordable Care Act," *CRS Report for Congress*, July 8, 2010, p.2, http://thf_media.s3.amazonaws.com/2010/pdf/R41315.pdf.

14. Edward Miller, "Health Reform Could Harm Medicaid Patients," *The Wall Street Journal*, December 4, 2009.
http://online.wsj.com/article/SB10001424052748703939404574567 98154918 4844.html?mod=djemEditorialPage.

15. Cynthia Tucker, "GOP's 'Lie of the Year' on Health Care Law," *The Atlanta Journal-Constitution*, January 4, 2011, http://blogs.ajc.com/cynthia-tucker/2011/01/04/gops-lie-of-the-year-on-health-care-law/.

16. Bill Adair and Angie Drobnic Holan, "PolitiFact's Lie of the Year: 'A Government Takeover of Health Care,'" December 16, 2010, http://politifact.com/truth-o-meter/article/2010/dec/16/lie-year-government-takeover-health-care/.

17. "200 Economists and Experts Support Repeal of Healthcare Law to Promote Job Growth, Reduce Deficit," American Action Forum, January 18, 2011, http://americanactionforum.org/sites/default/files/Final%20Open%20Letter_Impact%20of%20Healthcare%20Repeal_1182010.pdf.

18. "Paul Ryan on Health Care and the Steep Climb to Reclaim the American Idea," U.S. House of Representatives, floor debate on H.R. 3590 and H.R. 4872, March 21, 2010, www.budget.house.gov/News/DocumentSingle.aspx?DocumentID=209553.

What's in This Law: An Overview

1. Amy Chozik, "Obama Touts Single Payer System for Health Care," *The Wall Street Journal*, August 19, 2008, http://blogs.wsj.com/washwire/2008/08/19/obama-touts-single-payer-system.

2. Nancy Pelosi, "Remarks to the 2010 Legislative Conference for the National Association of Counties," March 9, 2010, www.speaker.gov/newsroom/pressreleases?id=1576.

3. Public Law 111–148 is divided into ten titles, which amend or reference a huge number of other statutes, including, as federal district court judge Henry Hudson noted, an estimated four hundred "provisions and riders patently extraneous to health care."

4. Representative Jeb Hensarling (R-Tex.), House floor speech, March 21, 2010, www.youtube.com/watch?v=IWcJun2DLXw.

5. The Gallup Poll, "Congress Only Growing Less Popular With Americans," September 20, 2010, www.gallup.com/poll/143054/congress-growing-less-popular-with-americans.aspx.

6. In 2011, estimated government funding will reach $1,357.7 billion and private funding will reach $1,352.1 billion. See Andrea M. Sisko, Christopher J. Truffer, Sean P. Keehan, John A. Poisal, M. Kent Clemens, and Andrew J. Madison, "National Health Spending Projections: The Estimated Impact of Reform Through 2019," *Health Affairs* (Web First) 29, no. 10, October 2010, http://rsc.jordan.house.gov/UploadedFiles/2010-0788_final.pdf, p. 2.

7. It is often asserted that the bills together amount to 2,700 pages, as stated by Judge Henry Hudson in his December 13, 2010, decision in *Commonwealth of Virginia v. Kathleen Sebelius, Secretary of the Department of Health and Human Services*. Our figure of 2,801 pages is based on an analysis of the legislation by the Minority Staff of the Joint Economic Committee of the Congress, www .jec.senate.gov/republicans/public/index.cfm?p=CommitteeNews&Content Record_id=bb302d88-3dOd-4424-8e33-3c5d2578c2bO.

Not surprisingly, the law is technically flawed. As *The Examiner* (Washington, D.C.) notes, "Thanks to simple drafting errors, their bill as written did not immediately require coverage of children with pre-existing conditions. It unexpectedly killed off a discount program for children's hospitals buying orphaned drugs. It imperiled workers' health care coverage at thousands of companies, forcing the Obama administration to waive the law's provisions for 222 employers (so far). It also lacked a severability clause—the legal provision ensuring that if a court strikes down part of a law, the remainder is preserved." "The ObamaCare Christmas Tree Gets Trimmed," editorial, *The Examiner*, December 14, 2010, p. 2.

8. Patient Protection and Affordable Care Act, Public Law 111–148, www .gpo.gov/fdsys/pkg/PLAW-111publ148/pdf/PLAW-111publ148.pdf.

For a summary of the combined legislation, see "Summary of New Health Reform Law," The Henry J. Kaiser Family Foundation, June 18, 2010, www.kff .org/healthreform/upload/8061.pdf.

9. There are several limited exemptions to this individual mandate, including ones for Native American Indian tribal members, members of a "recognized religious sect or division" with "established tenets or teaching" that would prohibit them from participating in public or private insurance, and members of nonprofit "health sharing ministries." There is also a more general "affordability exemption" applied to employees offered employer-based coverage or coverage in a state exchange if their out-of-pocket premium costs would exceed 8 percent of their household income. Other "hardship" exemptions would be determined on a case-by-case basis by the secretary of HHS. For further information, see the Congressional Research Service reports: The PPACA Penalty Provision and the Internal Revenue Service, http:// s3.amazonaws.com/thf_media/2010/pdf/Coburn-IRS-enforce-insurance-requirement.pdf; Private Health Insurance Provisions in PPACA (P.L. 111–148); also www.healthlawyers.org/Events/Programs/Materials/Documents/ HCR10/barnes_cowart_CRSprivatehealth.pdf.

10. This estimate is based on an analysis of the Senate version of the legislation by the staff of the Senate Republican Policy Committee.

11. Paul Starr, *The Social Transformation of American Medicine: The Rise of a Sovereign Profession and the Making of a Vast Industry* (New York: Basic Books, 1982), p. 235.

12. The Patient Protection and Affordable Care Act, Public Law 111–148, Section 1501.

13. Minority Staff Report, "The Wrong Prescription: Democrats' Health Overhaul Dangerously Expands IRS Authority," Prepared for Representative Dave Camp (R-Mich.) and Representative Charles Boustany (R-La.), House Committee on Ways and Means, March 18, 2010, http://republicans .waysandmeans.house.gov/UploadedFiles/IRS_Power_Report.pdf.

14. For the flat dollar amount, the annual penalty starts at $95 in 2014, rising to $325 in 2015 and $695 in 2016. For the percentage of income, it will start at 1 percent of income in 2014 and rise to 2 percent in 2015 and 2.5 percent in 2016. The penalty is to be assessed on all individuals the law deems eligible to be penalized, but there will be a family cap of $2,085 in 2016. In 2017, the annual increase in the amount of the penalty is to track the increase in inflation as measured by the Consumer Price Index (CPI).

15. The statutory language on this issue is tightly written and directly references Section 1402(g)(1) of the Internal Revenue Code, which notes that the sect must have been in continuous existence since December 31, 1950.

16. The Patient Protection and Affordable Care Act, Public Law 111–148, Section 1502.

17. This means that the employee works on average at least thirty hours a week.

18. The Patient Protection and Affordable Care Act, Public Law 111–148, Section 1513.

19. Ibid.

20. The Patient Protection and Affordable Care Act, Public Law 111–148, Section 1511.

21. The Patient Protection and Affordable Care Act, Public Law 111–148, Section 1514.

22. U.S. Chamber of Commerce, "Critical Employer Issues in the Patient Protection and Affordable Care Act" (Washington, D.C., 2010), p. 25.

23. One more item is worth noting. For employers who provide coverage for persons over age fifty-five who are *not* eligible for Medicare, the new law creates a new federal reinsurance program, with $5 billion allocated to the purpose, to offset 80 percent of the costs of health care claims between $15,000 and $90,000 in any given year for those persons.

24. Philip Bredesen, "ObamaCare's Incentive to Drop Insurance," *The Wall Street Journal*, October 21, 2010, http://online.wsj.com/article/SB100014240527 0230451070457556264380401525 2.html.

25. Medicaid is jointly financed by the federal government and the states. The federal government pays an average of about 57 percent of Medicaid costs, but the percentage is much higher in some states with a higher percentage of lower-income citizens. While Medicaid initially focused on providing care for poor young families (especially single mothers and their babies), much more of its funding is now channeled into long-term care in nursing homes and other forms of custodial care for the elderly and disabled.

26. These are 2010 dollar values. In the reconciliation bill (Public Law 111–152), Congress modified the eligibility standard from 133 percent to 138 percent of the federal poverty level by "disregarding" additional income of up to 5 percent of FPL.

27. Sisko et al., "National Health Spending Projections: The Estimated Impact of Reform Through 2019," p. 2.

28. These examples are taken from Douglas Holtz-Eakin, "Statement for the Forum to Examine the Status of the Implementation of the PPACA, Senate Republican Policy Committee," September 23, 2010, p. 6.

29. Letter to Senator Evan Bayh from CBO Director Douglas Elmendorf on the premium effects of the Patient Protection and Affordable Care Act, November 30, 2009, www.cbo.gov/ftpdocs/107xx/doc10781/11-30-Premiums.pdf.

30. In other words, the structure of ObamaCare's taxpayer subsidies for insurance premiums in exchanges is radically different from the voucher or tax-credit approaches that are commonly proposed by conservative or centrist health care policy analysts. We aim at establishing health care financing mechanisms that are predictable, fully transparent, and designed to help consumers to see health care costs and increase their range of competitive choices. This would provide strong incentives to secure value for money and economize in health care spending.

31. The exchange subsidies keep out-of-pocket health insurance premium payments within a specified percentage of income. For example, they limit those payments to 2 percent of income for lower-income persons below 133 percent of the federal poverty level and to no more than 9.5 percent of income for people at 400 percent of the poverty level ($88,200). The annual growth in the taxpayer premium subsidies will initially be tied to the annual average growth in premiums.

32. Letter from CBO Director Douglas W. Elmendorf to Rep. John Dingell, November 20, 2009, www.cbo.gov/ftpdocs/107xx/doc10741/hr3962Revised.pdf.

33. The Patient Protection and Affordable Care Act, Public Law 111–148, Section 8002(a).

34. The CLASS program is scheduled to begin in 2011. In its initial assessment, the Congressional Budget Office has estimated that the premium revenues from CLASS over the period 2010 to 2019 will amount to $70.2 billion.

35. Richard S. Foster, Chief Actuary, "Estimated Financial Effects of the

'Patient Protection and Affordable Care Act', as Amended," U.S. Department of Health and Human Services, Centers for Medicare & Medicaid Services, Office of the Actuary, April 22, 2010, www.cms.gov/ActuarialStudies/Downloads/PPACA_2010-04-22.pdf, p. 21.

36. Lori Montgomery, "Proposed Long-Term Insurance Program Raises Questions," *The Washington Post*, October 27, 2009, www.washingtonpost.com/wp-dyn/content/article/2009/10/27/AR2009102701417.html.

37. Foster, "Estimated Financial Effects," p. 2.

38. The Patient Protection and Affordable Care Act, Public Law 111–148, Section 3002.

39. Foster, "Estimated Financial Effects," p. 2.

40. Estimated Revenue Effects Of The Amendment In The Nature Of A Substitute To H.R. 4872, The "Reconciliation Act Of 2010," As Amended, In Combination With The Revenue Effects Of H.R. 3590, The "Patient Protection And Affordable Care Act ('PPACA')," As Passed By The Senate, And Scheduled For Consideration By The House Committee On Rules On March 20, 2010, March 20, 2010, www.jct.gov/publications.html?func=startdown&id=3672.

41. In 2011, if you use funds from your health savings account to purchase goods or services that are disallowed under the law, you will face an increased tax penalty of 20 percent (up from the previous 10 percent). In addition, you will find new limits on the number of products, such as over-the-counter medications, you can purchase with funds from tax-free health savings accounts or flexible spending accounts. There will also be a new tax on brand-name drugs in 2011. Expect the tax on drugs, like other new taxes on medical devices and insurance, to be passed on to you in the form of higher health care costs or insurance premiums.

42. Foster, "Estimated Financial Effects," p. 5.

43. Under Title I, state officials are instructed that they "shall" establish an "American Health Benefit Exchange" (AHB) in each state. The secretary is authorized to make grants to states to set up the exchanges. He or she will also have a lot of latitude in determining the amount of the money and whether or not to renew the grant, depending on whether or not your state is "making progress" in establishing the exchange. Each state must also meet the new federal insurance requirements and other "benchmarks" the secretary may see fit to establish. The secretary also will decide whether to certify proposed state health care benefit exchanges as "qualified" or reject them, as of January 1, 2013.

44. The Patient Protection and Affordable Care Act, Public Law 111–148, Section 1321(c)(1).

45. If the secretary approves, state officials may also enter into an agreement or compact with other states and establish multistate exchanges, but only if they meet or exceed the statutory and regulatory requirements under the new health care law.

46. The Patient Protection and Affordable Care Act, Public Law 111–148, Section 1311(d)(4).

47. The Patient Protection and Affordable Care Act, Public Law 111–148, Section 1311(e)(1)(B).

48. The Patient Protection and Affordable Care Act, Public Law 111–148, Section 1321.

49. The Patient Protection and Affordable Care Act, Public Law 111–148, Section 1332.

50. The Patient Protection and Affordable Care Act, Public Law 111–148. Section 1334(d).

51. Smaller insurance companies would be unlikely to secure these government contracts. According to the Congressional Budget Office, "Whether insurers would be interested in offering such plans is unclear, and establishing a nationwide plan comprising only nonprofit insurers might be particularly difficult. Even if such plans were arranged, the insurers offering them would probably have participated in the insurance exchanges anyway, so the inclusion of this provision did not have a significant effect on the estimates of federal costs or enrollment in the exchanges." Letter from Douglas Elmendorf, Director, Congressional Budget Office, to the Hon. Harry Reid, Majority Leader, United States Senate, concerning the spending and revenue estimates of the Patient Protection and Affordable Care Act, December 19, 2009, p. 9, www.cbo.gov/ftpdocs/108xx/doc10868/12-19-Reid_Letter_Managers_Correction_Noted.pdf.

52. Henry J. Kaiser Family Foundation, "Summary of New Health Reform Law," last modified June 18, 2010, www.kff.org/healthreform/upload/8061.pdf, p. 4.

53. Stuart M. Butler, "Why The Health Reform Wars Have Only Just Begun," Heritage Lectures no. 1158, July 6, 2010, http://report.heritage.org/hl1158.

54. John S. Hoff, "Implementing ObamaCare: A New Exercise in Old-Fashioned Central Planning," Backgrounder 2459, Heritage Foundation, September 10, 2010, www.heritage.org/Research/Reports/2010/09/Implementing-Obamacare-A-New-Exercise-in-Old-Fashioned-Central-Planning.

55. These rules apply primarily to the individual market because group insurance plans have been subject to somewhat similar restrictions since 1996 under another federal law, called the Health Insurance Portability and Accountability Act.

56. The Patient Protection and Affordable Care Act, Public Law 111–148, Section 2711.

57. The out-of-pocket limits on each of the plans are equal to those allowable under the law governing health savings accounts ($5,950 for singles and $11,900 for families in 2010 dollars): the Bronze Plan covers 60 percent of the essential

benefit costs; the Silver Plan covers 70 percent of the essential benefit costs; the Gold Plan covers 80 percent of the essential benefit costs; and the Platinum Plan covers 90 percent of the essential benefit costs. There is also a special Catastrophic Plan available to you if you are thirty years of age or younger and you are exempt from the federal mandate to purchase health insurance.

58. Proposed federal rules for such state-based rate reviews ran behind schedule, arriving for comment in December 2010, so they won't go into effect until later in 2011.

59. "Paul Ryan on Health Care and the Steep Climb to Reclaim the American Idea," U.S. House of Representatives, floor debate on H.R. 3590 and H.R. 4872, March 21, 2010, www.budget.house.gov/News/DocumentSingle .aspx?DocumentID=209553.

Impact On. . .Families and Young Adults

1. Senator Barack Obama, in a meeting with the *San Francisco Chronicle*'s editorial board, January 17, 2008, www.sfgate.com/cgi-bin/object/article?f=/ c/a/2008/01/19/EDIAUHASH.DTL.

2. Remarks of Senator Barack Obama, Ohio State University Medical Center, Saturday, February 23, 2008, Columbus, Ohio, http://nationalelections .blogspot.com/2008_02_01_archive.html.

3. Calvin Woodward and Erica Werner, "FACT CHECK: Obama's Tone Shifts on Health Care," *Associated Press*, September 11, 2010, http://abcnews .go.com/Politics/wireStory?id=11606276.

4. Congressional Budget Office and the Joint Committee on Taxation, "An Analysis of Health Insurance Premiums Under the Patient Protection and Affordable Care Act," November 30, 2009, www.cbo.gov/ftpdocs/107xx/ doc10781/11-30-Premiums.pdf.

5. "Durbin Admits Premiums Will Go Up If Health Care Bill Is Passed," Senate Republican Communications Center, March 10, 2010, www.youtube .com/watch?v=m7QAci-XWHY.

6. Michael Barone, "Gangster Government Stifles Criticism of Obama-Care," *The Examiner*, September 10, 2010, http://washingtonexaminer.com/ politics/gangster-government-stifles-criticism-ObamaCare.

7. "EXCLUSIVE: Vice President Biden Says President Obama's Rescheduled Trip Not a Bad Sign for Health Care Bill's Prospects," ABC News, March 18, 2010, http://blogs.abcnews.com/politicalpunch/2010/03/exclusive-vice-president-biden-says-obamas-cancelled-trip-not-a-bad-sign-for-health-care-bills-prosp.html.

8. "State by State Enrollment in the Pre-Existing Condition Insurance Plan, as of November 1, 2010," November 5, 2010, at www.healthcare.gov/news/ factsheets/pre-existing_condition_insurance_enrollment.html.

9. Amy Goldstein, "Health Plans for High-Risk Patients Attracting Fewer, Costing More than Expected," *The Washington Post*, December 27, 2010, www .washingtonpost.com/wp-dyn/content/article/2010/12/27/AR2010122702343 .html?sid=ST2010122702477.

10. John Reichard, "Analyst: Health Law Means Upheaval for Health Care Economy," *CQ HealthBeat News*, November 11, 2010, www.cq.com/doc/ hbnews-3761576?wr=RDlYTlRja3lSajRoWWFoZGNwN084UQ.

11. Henry J. Kaiser Family Foundation, "Summary of New Health Reform Law," June 18, 2010, www.kff.org/healthreform/upload/8061.pdf.

12. Richard S. Foster, "Estimated Financial Effects of the 'Patient Protection and Affordable Care Act,' as Amended," Office of the Actuary, Centers for Medicare & Medicaid Services, April 22, 2010, www.cms.gov/ActuarialStudies/ Downloads/PPACA_2010-04-22.pdf.

13. Reed Abelson, "Insurer Cuts Health Plans as New Law Takes Hold," *The New York Times*, September 30, 2010, www.nytimes.com/2010/10/01/health/ policy/01insure.html.

14. "Under the law, adults under age 30 who are not eligible for subsidized coverage will have the option to purchase a catastrophic health plan. The plan will be required to have the essential benefits package and include three primary care visits per year but could have cost-sharing similar to health savings account-eligible, high-deductible plans. . . . Preventive services will be excluded from the deductible and cost-sharing would be limited to the current health savings account out-of-pocket limits ($5,950 for single policies and $11,200 for families). People over age 30 who could not find a plan with a premium that is 8 percent or less of their income would be able to purchase the young adult plan, as well." Sara R. Collins and Jennifer L. Nicholson, "Realizing Health Reform's Potential: Young Adults and the Affordable Care Act of 2010," October 2010, www.commonwealthfund.org/~/media/Files/Publications/ Issue%20Brief/2010/Oct/1446_Collins_young_adults_and_ACA_ib.pdf.

15. Kathleen Sebelius, "Health Insurers Finally Get Some Oversight," *The Wall Street Journal*, September 28, 2010, http://online.wsj.com/article/SB100014 24052748704082104575515851336184716.html.

16. Diana Furchtgott-Roth, "Healthcare's Impact on the Low-Skilled Worker," May 6, 2010, www.realclearmarkets.com/articles/2010/05/06/healthcare_ and_low-skilled_workers_98451.html.

17. Diana Furchtgott-Roth, "ObamaCare Will Hurt Low-Skill Workers," *The Examiner*, December 16, 2010, http://washingtonexaminer.com/opinion/ columnists/2010/12/diana-furchtgott-roth-ObamaCare-will-hurt-low-skill- workers.

18. "Boehner Files Amicus Brief Challenging Constitutionality of Individual Mandate at Heart of Job-Killing ObamaCare Law," November 16, 2010, http:// johnboehner.house.gov/News/DocumentSingle.aspx?DocumentID=215122.

19. The health care law uses a complicated formula to decide how much employers must pay. Employers with more than fifty employees that do not offer coverage and have at least one full-time employee who receives a premium tax credit will be assessed $2,000 per full-time employee, excluding the first thirty employees from the assessment. Employers with more than fifty employees that offer coverage but have at least one full-time employee receiving a premium tax credit will pay the lesser of $3,000 for each employee receiving a premium credit or $2,000 for each full-time employee. Henry J. Kaiser Family Foundation, "Summary of New Health Reform Law."

20. Senate Budget Committee Republican Staff, "Budget Perspective: The Real Deficit Effect of the Health Bill," December 22, 2009. This estimate was made on the Senate version of ObamaCare, not the final bill. The Senate bill is close to what was passed in final form; therefore, $2.3 trillion is likely to be an accurate ten-year estimate of spending in the final bill as well.

21. PPACA provides for refundable and advanceable premium credits to eligible individuals and families with incomes from 133 to 400 percent of the Federal Poverty Level (FPL) to purchase insurance through the exchanges. The premium credits will be tied to the second-lowest-cost silver plan in the area and will be set on a sliding scale such that the premium contributions are limited to the following percentages of income:

HOUSEHOLD INCOME	PREMIUM CONTRIBUTION LIMIT
Up to 133% FPL:	2% of income
133–150% FPL:	3–4% of income
150–200% FPL:	4–6.3% of income
200–250% FPL:	6.3–8.05% of income
250–300% FPL:	8.05–9.5% of income
300–400% FPL:	9.5% of income

Henry J. Kaiser Family Foundation, "Summary of New Health Reform Law."

22. "Obama Flip-Flops on Requiring People to Buy Health Care," *St. Petersburg Times*, July 20, 2009, www.politifact.com/truth-o-meter/statements/2009/jul/20/barack-obama/obama-flip-flops-requiring-people-buy-health-care/.

23. "Transcript: Reps. Ryan, Wasserman Schultz on 'FNS,'" *FOX News Sunday Interview*, March 21, 2010, www.foxnews.com/story/0,2933,589726,00.html.

24. Foley & Lardner LLP, "Patient Protection and Affordable Care Act (PPACA): Employment-Related Provisions," 2010, www.foley.com/files/PPACASummary.pdf.

25. Paul Ryan, "Costs of This Debacle Will Be High," *Milwaukee Journal Sentinel*, March 23, 2010, www.jsonline.com/news/opinion/88960777.html.

Impact on . . . Seniors

1. "2010 Annual Report of the Boards of Trustees of the Federal Hospital Insurance and Federal Supplementary Medical Insurance Trust Funds," The Boards of Trustees, Federal Hospital Insurance and Federal Supplementary Medical Insurance Trust Funds, August 2010, https://www.cms.gov/ReportsTrustFunds/downloads/tr2010.pdf.

2. Elliott S. Fisher, Mark B. McClellan, John Bertko, Steven M. Lieberman, Julie J. Lee, Julie L. Lewis, and Jonathan S. Skinner, "Fostering Accountable Health Care: Moving Forward in Medicare," *Health Affairs* (Web exclusive), January 27, 2009, http://content.healthaffairs.org/content/28/2/w219.abstract.

3. The relevant section is the new Section 1899(c) of the Social Security Act, as created in Section 3022 of the Patient Protection and Affordable Care Act, as Amended.

4. Section 3403, as amended, of the PPACA. Beginning in 2015, per capita Medicare spending is supposed to be limited to a fixed growth rate, initially set according to a mix of general inflation in the economy and inflation in the health sector. Starting in 2018, the upper limit is set permanently at per capita gross domestic product growth plus one percentage point.

5. Letter to House Speaker Nancy Pelosi from CBO Director Douglas Elmendorf on the cost estimates for the Patient Protection and Affordable Care Act, March 20, 2010, www.cbo.gov/ftpdocs/113xx/doc11379/AmendReconProp.pdf.

6. Solomon M. Mussey, Office of the Actuary, Centers for Medicare & Medicaid Services, "Estimated Effects of the 'Patient Protection and Affordable Care Act,' as Amended, on the Year of Exhaustion for the Part A Trust Fund, Part B Premiums, and Part A and Part B Coinsurance Amounts," memorandum, April 22, 2010, www.cms.gov/ActuarialStudies/Downloads/PPACA_Medicare_2010-04-22.pdf.

7. "Effects of the Patient Protection and Affordable Care Act on the Federal Budget and the Balance in the Hospital Insurance Trust Fund," Congressional Budget Office, December 23, 2009, http://cboblog.cbo.gov/?p=448.

8. *2010 Annual Report of the Boards of Trustees of the Federal Hospital Insurance and Federal Supplementary Medical Insurance Trust Funds*, August 2010, www.cms.gov/ReportsTrustFunds/downloads/tr2010.pdf, pp. 281–283.

9. Douglas Holtz-Eakin, Joseph Antos, and James C. Capretta, "Health Care Repeal Won't Add to the Deficit," *The Wall Street Journal*, January 19, 2011, http://online.wsj.com/article/SB1000142405274870395400457608970235429210000.html.

10. John D. Shatto and M. Kent Clemens, Office of the Actuary, Centers for Medicare & Medicaid Services, U.S. Department of Health and Human Services, "Projected Medicare Expenditures Under an Illustrative Scenario with Alternative Payment Updates to Medicare Providers," memorandum, August 5, 2010, https://www.cms.gov/ReportsTrustFunds/downloads/2010TRAlternativeScenario.pdf, p. 6.

11. Ibid.

12. Kaiser Family Foundation, "Medicare Advantage Fact Sheet," September 2010, www.kff.org/medicare/upload/2052-14.pdf.

13. Medicare Payment Advisory Commission, *Healthcare Spending and the Medicare Program: A Data Book*, June 2010, www.medpac.gov/documents/Jun10DataBookEntireReport.pdf, p. 162.

14. National Bipartisan Commission on the Future of Medicare, "Final Breaux-Thomas Medicare Proposal," March 16, 1999, http://thomas.loc.gov/medicare/bbmtt31599.html.

15. For instance, if a Medicare beneficiary does not make a conscious choice about his or her health coverage, he or she ends up in the traditional Medicare program, not a Medicare Advantage plan. The government therefore doesn't really need to advertise to get enrollment into its program. That's not true for Medicare Advantage.

In addition, many of the administrative costs of running the traditional Medicare program, such as revenue collection by the Internal Revenue Service, are not factored into the premium paid by beneficiaries. Most important, the traditional program dictates the prices it will pay to medical service providers. Private Medicare Advantage plans have no such option. They must negotiate contracts with their networks and get hospitals and physicians to agree to a fee schedule. Moreover, in many parts of the country, private plans are forced to pay premium rates to compensate hospitals and physicians for the losses they incur on their Medicare fee-for-service business.

16. Medicare Payment Advisory Commission, *Healthcare Spending and the Medicare Program: A Data Book*, p. 162.

17. Congressional Budget Office cost estimate for the PPACA, as amended by H.R. 4872, March 20, 2010, www.cbo.gov/ftpdocs/113xx/doc11379/AmendReconProp.pdf.

18. Richard S. Foster, "Estimated Financial Effects of the 'Patient Protection and Affordable Care Act,' as Amended," Centers for Medicare & Medicaid Services, Office of the Actuary, April 22, 2010, www.cms.gov/ActuarialStudies/Downloads/PPACA_2010-04-22.pdfPPACA_2010-04-22.pdf.

19. This cut comes from the combined effect of changing the formula for paying Medicare Advantage plans and the pass-through of cuts in what traditional Medicare pays to hospitals and other institutions.

20. Robert A. Book and James C. Capretta, "Reductions in Medicare

Advantage Payments: The Impact on Seniors by Region," Backgrounder 2464, Heritage Foundation, September 14, 2010, www.heritage.org/research/reports/2010/09/reductions-in-medicare-advantage-payments-the-impact-on-seniors-by-region#_ftn3.

21. Heritage Foundation, "County-Level Effects of Medicare Advantage Changes in the Patient Protection and Affordable Care Act (PPACA)," September 2010, http://thf_media.s3.amazonaws.com/2010/pdf/MA_County_Results_Summary.pdf .

22. Book and Capretta, "Reductions in Medicare Advantage Payments."

23. Ibid.

24. Ibid.

25. CBO Letter to House Speaker Nancy Pelosi, March 20, 2010, www.cbo.gov/ftpdocs/113xx/doc11379/AmendReconProp.pdf.

Impact on. . .Vulnerable Americans

1. Kaiser Commission on Medicaid and the Uninsured and Urban Institute estimates based on 2007 MSIS and CMS64 data. Henry J. Kaiser Family Foundation, "The Medicaid Program at a Glance," June 2010, www.kff.org/medicaid/upload/7235-04.pdf.

2. "Medicare/Medicaid Acceptance Trends Among Physicians," June 2010, http://amerimedconsulting.com/home/wp-content/uploads/press/pdf/AmeriMedTrendTracker-2010.pdf.

3. Tamyra Carroll Garcia, Amy B. Bernstein, and Mary Ann Bush, "Emergency Department Visitors and Visits: Who Used the Emergency Room in 2007?," NCHS Data Brief no. 38, National Center for Health Statistics, Centers for Disease Control and Prevention, May 2010, www.cdc.gov/nchs/data/databriefs/db38.pdf.

4. Donna Cohen Ross, Marian Jarlenski, Samantha Artiga, and Caryn Marks, "A Foundation for Health Reform: Findings of a 50 State Survey of Eligibility Rules, Enrollment and Renewal Procedures, and Cost-Sharing Practices in Medicaid and CHIP for Children and Parents During 2009," Kaiser Commission on Medicaid and the Uninsured, Henry J. Kaiser Family Foundation, December 2009, www.kff.org/medicaid/8028.cfm.

5. Henry J. Kaiser Family Foundation, "Summary of New Health Reform Law," June 18, 2010, www.kff.org/healthreform/upload/8061.pdf.

6. Clifford J. Levy and Michael Luo, "New York Medicaid Fraud May Reach into Billions," The New York Times, July 18, 2005, www.nytimes.com/2005/07/18/nyregion/18medicaid.html?_r=2.

7. Michael Luo and Clifford J. Levy, "As Medicaid Balloons, Watchdog Force Shrinks," The New York Times, July 19, 2005, www.nytimes.com/2005/07/19/nyregion/19medicaid.html.

8. Debra Beaulieu, "New York Medicaid Fraud Collections, Convictions Hit Record High," April 13, 2010, www.fiercehealthcare.com/story/new-york-medicaid-fraud-collections-convictions-hit-record-high/2010-04-13.

9. Edward Miller, "Health Reform Could Harm Medicaid Patients," *The Wall Street Journal*, December 4, 2009, http://online.wsj.com/article/SB1000142 4052748703939404574567981549184844.html?mod=djemEditorialPage.

10. Former governor Mitt Romney, a Republican, led the Massachusetts reform effort. His two leading initiatives were to create a health insurance exchange that would allow employees of small businesses to select the private health care coverage of their choice. Employers could contribute to the coverage, and employees could pay premiums with pretax dollars. Importantly, the insurance would be portable so people wouldn't lose it as they moved from job to job. Second, Romney redirected subsidies to hospitals and other institutions for treating the uninsured to help people get insurance instead.

But in the political process, both objectives were blunted. The exchange became the "Commonwealth Connector," which emerged as a powerful regulatory agency. It required plans to cover a long list of health care benefits and limited the diversity and number of plans that could be offered. The interests of small businesses were subordinated to other priorities, and the only options initially offered to people getting subsidized insurance were Medicaid managed care plans. For a description of the implementation of the Massachusetts law and its impact on small businesses, see Joshua D. Archambault, "Massachusetts Health Care Reform Has Left Small Business Behind: A Warning to the States," Heritage Foundation Backgrounder no. 2462, September 16, 2010, http://report.heritage.org/bg2462.

11. When Romney signed health care reform legislation in 2006, he vetoed certain provisions, including the employer mandate, which were overridden by the legislature. The final law included components that are similar to the provisions of ObamaCare, including a major expansion of Medicaid, a "connector" with powers to mandate health care benefits, a mandate that most employers provide insurance or pay a fine, and an individual mandate that residents of the Commonwealth must carry health insurance or pay a fine.

12. Michael Norton and Kyle Cheney, "Medicaid Costs Surge Past $10 Billion, Devouring Uptick in Tax Receipts," *State House News Service*, December 13, 2010, www.statehousenews.com/cgi/as_web.exe?rev2010+D+15040366.

13. Ibid.

14. Michael Levenson, "Health Costs Sap State Aid for Schools," *The Boston Globe*, December 9, 2010, www.boston.com/news/education/k_12/articles/2010/12/09/health_care_costs_sap_aid_for_massachusetts_schools/.

15. Richard S. Foster, "Estimated Financial Effects of the 'Patient Protection and Affordable Care Act,' as Amended," Office of the Actuary, Centers for

Medicare & Medicaid Services, April 22, 2010, www.cms.gov/ActuarialStudies/Downloads/PPACA_2010-04-22.pdf.

16. Ibid. (The Congressional Budget Office puts the ten-year saving at $136 billion. Congressional Budget Office cost estimate for the PPACA, as amended by H.R. 4872, March 20, 2010, www.cbo.gov/ftpdocs/113xx/doc11379/AmendReconProp.pdf.)

17. Robert Weisman, "Harvard Pilgrim Cancels Medicare Advantage Plan," *The Boston Globe*, September 28, 2010, www.boston.com/business/healthcare/articles/2010/09/28/harvard_pilgrim_cancels_medicare_advantage_plan/; Harvard Pilgrim Health Care, "First Seniority Freedom," https://www.harvardpilgrim.org/portal/page?_pageid=253,240302&_dad=portal&_schema=PORTAL.

18. Scott Gottlieb, "Health Reform's Competition Crushers," *New York Post*, November 23, 2010, www.aei.org/article/102811.

19. On March 21, 2010, the House passed the Senate version of health insurance reform legislation, the Patient Protection and Affordable Care Act (H.R. 3590), by a vote of 219 to 212—sending it to the president who signed it into law on March 23. The same day, the House also passed the Health Care Education Affordability Reconciliation Act (H.R. 4872), which amended several key provisions of the first bill, by a vote of 220 to 211; the Senate passed it on March 25, with two small changes, and it returned to the House later that evening, passing by a vote of 220 to 207. The president signed this bill on March 30, 2010. http://waysandmeans.house.gov/press/PRArticle.aspx?NewsID=11085.

20. "Remarks by the President in Health Insurance Reform Town Hall, Portsmouth High School, Portsmouth, New Hampshire," Office of the Press Secretary, The White House, August 11, 2009, www.whitehouse.gov/the-press-office/remarks-president-town-hall-health-insurance-reform-portsmouth-new-hampshire.

21. Robert Pear, "Obama Returns to End-of-Life Plan That Caused Stir," *The New York Times*, December 25, 2010, www.nytimes.com/2010/12/26/us/politics/26death.html?_r=1&nl=todaysheadlines&emc=a2.

22. Ibid.

23. Robert Pear, "U.S. Alters Rule on Paying for End-of-Life Planning," *The New York Times*, January 4, 2011, www.nytimes.com/2011/01/05/health/policy/05health.html.

24. "Background Questions and Answers on Health Care Plan," www.barackobama.com/pdf/Obama08_HealthcareFAQ.pdf.

25. "Transcript: ABC News 'Dr. Timothy Johnson Interviews President Barack Obama,'" ABC News, July 15, 2009, http://abcnews.go.com/Politics/story?id=8091227&page=1.

26. PPACA established the Patient Centered Outcomes Research Institute (PCORI), which replaced the CER Council.

27. U.S. Food and Drug Administration, "FDA Begins Process to Remove Breast Cancer Indication from Avastin Label," December 16, 2010, www.fda .gov/NewsEvents/Newsroom/PressAnnouncements/ucm237172.htm; Alicia Mundy, "Roche Drug Faces FDA Curb," *The Wall Street Journal*, December 17, 2010, http://online.wsj.com/article/SB10001424052748704073804576023512609 387864.html?KEYWORDS=avastin.

28. "Cleveland Clinic's Dr. Cosgrove on Healthcare Reform," August 18, 2010, http://my.clevelandclinic.org/news/2010/healthcare_reform_cosgrove .aspx.

29. National Institutes for Health and Clinical Excellence, "Measuring Effectiveness and Cost Effectiveness: the QALY," April 20, 2010, www.nice.org. uk/newsroom/features/measuringeffectivenessandcosteffectivenesstheqaly.jsp.

30. Valentin Petkantchin, "Risks and Regulatory Obstacles for Innovating Companies in Europe," Institut économique Molinari, October 2008, www .institutmolinari.org/IMG/pdf/cahier1008_en-2.pdf.

31. Michael Schlander, *Health Technology Assessments by the National Institute for Health and Clinical Excellence: A Qualitative Study* (New York: Springer, 2007); Michael Schlander, "Comparative Effectiveness Programs: A Global Perspective: Discussing Germany and the UK," Galen Institute, March 9, 2009, www.galen.org/fileuploads/SchlanderCEP.pdf.

32. Presentation on a panel about the health care outlook for the 112th Congress, December 15, 2010.

33. House Committee on Appropriations Report to Accompany H.R. 679, The American Recovery and Reinvestment Act of 2009, January 26, 2009, http://frwebgate.access.gpo.gov/cgi-bin/getdoc.cgi?dbname=111_cong_ reports&docid=f:hr004.111.pdf.

34. Tom Daschle with Jeanne M. Lambrew and Scott S. Greenberger, *Critical: What We Can Do About the Health-Care Crisis* (New York: Thomas Dunne Books, 2008).

35. Michel P. Coleman, Manuela Quaresma, Franco Berrino, Jean-Michel Lutz, Roberta De Angelis, Riccardo Capocaccia, Paolo Baili, Bernard Rachet, Gemma Gatta, Timo Hakulinen, Andrea Micheli, Milena Sant, Hannah K. Weir, J. Mark Elwood, Hideaki Tsukuma, Sergio Koifman, Gulnar Azevedo e Silva, Silvia Francisci, Mariano Santaquilani, Arduino Verdecchia, Hans H. Storm, John L. Young, and the CONCORD Working Group, "Cancer Survival in Five Continents: A Worldwide Population-Based Study (CONCORD)," *The Lancet Oncology* 9, no. 8 (August 2008), http://v1.theglobeandmail.com/v5/ content/pdf/CONCORD.pdf, pp. 730–756.

36. Nick Triggle, "Lucentis: An NHS dilemma," August 27, 2008, http:// news.bbc.co.uk/2/hi/health/7582740.stm.

37. Eric Lichtblau and Robert Pear, "Washington Rule Makers Out of the Shadows," *The New York Times*, December 8, 2010, www.nytimes .com/2010/12/09/us/politics/09rules.html?_r=1.

38. Curtis W. Copeland, "Initial Final Rules Implementing the Patient Protection and Affordable Care Act," Congressional Research Service, December 10, 2010, http://thehill.com/images/stories/blogs/crsrules.pdf.

39. "Statement from the President on Sunshine Week," Office of the Press Secretary, The White House, March 16, 2010, www.whitehouse.gov/the-press-office/statement-president-sunshine-week.

40. Letter from Republican Senate Finance Committee members to Senator Max Baucus, July 14, 2010, http://finance.senate.gov/newsroom/ranking/release/?id=1b3a715b-0a5d-4b0d-975b-9489d983c409.

41. Donald M. Berwick, "A Transatlantic Review of the NHS at 60," NHS Live, July 1, 2008, www.wales.nhs.uk/sites3/page.cfm?orgid=781&pid=32953.

42. "Rethinking Comparative Effectiveness Research," An Interview with Dr. Donald Berwick, Biotechnology Healthcare, June 2009, www.ncbi.nlm.nih.gov/pmc/articles/PMC2799075/pdf/bth06_2p035.pdf.

43. Amy B. Monahan and Daniel Schwarcz, "Will Employers Undermine Health Care Reform by Dumping Sick Employees?," Virginia Law Review, forthcoming; Minnesota Legal Studies Research Paper no. 10-37, July 30, 2010, http://ssrn.com/abstract=1651308.

44. Comment by Daniel Schwarcz in response to a post on Marginal Revolution, November 29, 2010, www.marginalrevolution.com/marginal revolution/2010/11/ObamaCare-where-are-we.html#comments.

45. America's Health Insurance Plans, "January 2010 Census Shows 10 Million People Covered by HSA/High-Deductible Health Plans," May 2010, www.ahipresearch.org/pdfs/HSA2010.pdf.

46. Grace-Marie Turner, Testimony before the U.S. Senate Committee on Health, Education, Labor, & Pensions Hearing on "Protection from Unjustified Premiums," April 20, 2010, www.galen.org/fileuploads/TurnerHELP0420.pdf.

Impact on . . . You and Your Doctor

1. Remarks of President Barack Obama—As Prepared for Delivery to the American Medical Association, June 15, 2009, http://blogs.wsj.com/washwire/2009/06/15/obama-if-you-like-your-doctor-you-can-keep-your-doctor/.

2. John Reichard, "Analyst: Health Law Means Upheaval for Health Care Economy," CQ HealthBeat News, November 11, 2010, www.cq.com/doc/hbnews-3761576?wr=RDlYTlRja3lSajRoWWFoZGNwN084UQ.

3. Reed Abelson, "Insurer Cuts Health Plans as New Law Takes Hold," The New York Times, September 30, 2010, www.nytimes.com/2010/10/01/health/policy/01insure.html.

4. Richard S. Foster, "Estimated Financial Effects of the 'Patient Protection

and Affordable Care Act,' as Amended," Office of the Actuary, Centers for Medicare & Medicaid Services, April 22, 2010, https://www.cms.gov/ActuarialStudies/Downloads/PPACA_2010-04-22.pdf.

5. Physicians Foundation, "2010 Survey: Physicians and Health Reform," 2010, at www.physiciansfoundation.org.

6. The Patient Protection and Affordable Care Act (Public Law 111–148) was signed into law on March 23, 2010. It was amended by the Health Care and Education Reconciliation Act of 2010 (Public Law 111–152), enacted on March 30, 2010. For convenience, the final legislation is referred to as the Patient Protection and Affordable Care Act (PPACA).

7. "Florida Doctor Stands By Anti-'Obamacare' Sign Despite Threat of Complaint," April 5, 2010, www.foxnews.com/politics/2010/04/05/florida-doctor-stands-anti-obamacare-sign-despite-threat-complaint/#ixzz1AOC5o2b6.

8. "The 2011 National Physicians: Frustration and Dismay in a Time of Change," Thomson Reuters and HCPlexus, January 18, 2011, http://hcplexus.com/survey.

9. Daniel Palestrant, "Why Physicians Oppose the Health Care Reform Bill," *Forbes*, April 28, 2010, www.forbes.com/2010/04/28/health-care-reform-physicians-opinions-contributors-daniel-palestrant.html.

10. The Physicians Foundation, "2010 Survey: Physicians and Health Reform."

11. Anna Wilde Mathews, "When the Doctor Has a Boss," *The Wall Street Journal*, November 8, 2010, http://online.wsj.com/article/SB10001424052748703865045756004127166683130.html.

12. Shatto and Clemens, "Projected Medicare Expenditures Under an Illustrative Scenario with Alternative Payment Updates to Medicare Providers."

13. The Patient Protection and Affordable Care Act, Public Law 111–148, Section 1311.

14. Ibid.

15. See National Committee for Quality Assurance, *The State of Health Care Quality: Reform, The Quality Agenda and Resource Use*, 2010, http://www.ncqa.org/portals/0/state%20of%20health%20care/2010/SOHC%202010%20-%20Full2.pdf, p. 39.

16. In 2015, physicians will receive 98.5 percent of reimbursements for failure to comply, and starting in 2016 and in each subsequent year, noncompliant physicians will receive 98 percent of reimbursements.

17. The Patient Protection and Affordable Care Act, Public Law 111–148, Section 3007.

18. The Patient Protection and Affordable Care Act, Public Law 111–148, Section 1303.

19. Gregory C. Kane, Michael R. Grever, John I. Kennedy, Mary Ann

Kuzma, Alan R. Saltzman, Peter H. Wiernik, and Nicole V. Baptista, "The Anticipated Physician Shortage: Meeting the Nation's Need for Physician Services," *The American Journal of Medicine* 122, no. 12 (December 2009), www .im.org/Publications/Ap.m.Perspectives/Documents/Dec09Perspectives.pdf, pp. 1156–1162.

20. Association of American Medical Colleges, "The Complexities of Physician Supply and Demand: Projections Through 2025," November 2008, http://services.aamc.org/publications/showfile.cfm?file=version122.pdf.

21. Kevin B. O'Reilly, "Health Reform's Next Challenge: Who Will Care for the Newly Insured?," April 12, 2010, www.ama-assn.org/amednews/2010/04/12/ prl10412.htm.

22. Roger E. Meyer, M.D., "Not Enough Doctors? Too Many? Why States Not Washington Must Solve the Problem," Heritage Foundation Backgrounder #2493, November 29, 2010, at http://report.heritage.org/bg2493.

23. Eric Helland and Mark H. Showalter, "The Impact of Liability on the Physician Labor Market," *The Journal of Law & Economics* 52 (November 2009), www.journals.uchicago.edu/doi/full/10.1086/597427.

24. Medicus Firm, "The Medicus Firm Physician Survey: Health Reform May Lead to Significant Reduction in Physician Workforce," January 2010, www .themedicusfirm.com/pages/medicus-media-survey-reveals-impact-health-reform.

25. After four years of undergraduate studies, future doctors must complete four years of medical school. Upon graduation from medical school, they must then complete an intense residency training program before they can practice independently. Depending on the specialty, residencies range from three to six years. After residency, most doctors also complete a fellowship. Fellowships typically involve an additional two to three years of training.

26. American Medical Association, "Medical Student Debt," www.ama-assn.org/ama/pub/about-ama/our-people/member-groups-sections/medical-student-section/advocacy-policy/medical-student-debt.shtml.

27. Under the Balanced Budget Act of 1997, Congress enacted a unique statutory restriction on the ability of doctors and Medicare patients to contract privately with each other for the delivery of medical services *outside* the program. Under that provision, doctors could privately contract if they notified the HHS secretary of their intent to enter into such a contract, submit the notification within ten days, and agree to forgo all other Medicare reimbursement from all other Medicare patients for a period of two years. No such restriction has ever been applied to any other government health program before or since, including Medicaid.

28. Patricia A. Davis, Jim Hahn, Paulette C. Morgan, Julie Stone, and Sibyl Tilson, "The Medicare Provisions In PPACA (P.L.111–148)," April 21, 2010, http://healthreform.kff.org/~/media/Files/KHS/docfinder/crsmedicareprovisions .pdf, p. 8.

29. Richard Wolf, "Doctors limit new Medicare patients," *USA Today*, June 21, 2010, www.usatoday.com/news/washington/2010-06-20-medicare_N.htm.

30. John G. R. Howie, David J. Heaney, Margaret Maxwell, Jeremy J. Walker, George K. Freeman, and Harbinder Rai, "Quality at General Practice Consultations: Cross Sectional Survey," *British Medical Journal* 319 (September 18, 2009), www.bmj.com/content/319/7212/738.full, pp. 738–743.

31. The Patient Protection and Affordable Care Act, Public Law 111–148, Section 6301.

32. Stuart MacDonald Butler, "Concerns Presented by the New Health Legislation," *The Cancer Journal* 16, no. 6 (November–December 2010), http://journals.lww.com/journalppo/Fulltext/2010/11000/Concerns_Presented_by_the_New_Health_Legislation.13.aspx, p. 3.

33. Congressional Budget Office, "Research on the Comparative Effectiveness of Medical Treatments," December 2007, www.cbo.gov/ftpdocs/88xx/doc8891/12-18-ComparativeEffectiveness.pdf, p. 31.

34. "Rethinking Comparative Effectiveness Research," *Biotechnology Healthcare* 6, no. 2 (June 2009), www.ncbi.nlm.nih.gov/pmc/articles/PMC2799075/, pp. 35–36, 38.

35. Jane M. Orient, M.D., *Your Doctor Is Not In* (New York: Crown Publishers, 1994), p. 66.

Impact on. . .You and Your Employer

1. In 2009, approximately 170 million Americans received private health insurance coverage through an employer—either directly as employees participating in an employer health plan or as dependents of a worker with job-based health care benefits. They represent almost 57 percent of all Americans, a decline from a recent high mark of 64 percent in 2000. Carmen DeNavas-Walt, Bernadette D. Proctor, and Jessica C. Smith, U.S. Census Bureau, *Income, Poverty, and Health Insurance Coverage in the United States: 2009*, Current Population Reports, P60-238 (Washington, D.C.: U.S. Government Printing Office, 2010), table C-1.

2. Modern forms of health insurance began to develop in the 1930s, but they gained their biggest boost from the special tax treatment of employer-sponsored insurance (ESI) that began during World War II and was secured in permanent statutory law in 1954. The decades of the 1950s and 1960s, in particular, experienced substantial growth of ESI as the dominant form of private health insurance. For a more detailed history of employer-sponsored health insurance and tax policy on health care, see Robert B. Helms, "Tax Policy and the History of the Health Insurance Industry," in *Using Taxes to Reform Health Insurance*, ed. Henry J. Aaron and Leonard E. Burman (Washington, D.C.: Brookings Institution Press, 2008), pp. 13–35; and Melissa A. Thomasson, "The Importance of Group Coverage: How Tax Policy Shaped U.S. Health Insurance," *American*

Economic Review 93, no. 4 (September 2003), pp. 1373–1384. Both reviews of the history of private health insurance found that over the three decades following World War II, coverage of people in employer-paid groups expanded relative to coverage of people who bought individual, nongroup policies.

3. In addition, neither employer-paid insurance premiums nor health care benefits financed directly by self-insured employers are counted as employee wages for purposes of the payroll taxes that finance Social Security and Medicare.

4. Purchases of some other forms of private health insurance do receive more limited tax benefits. Self-employed workers can claim an income tax deduction for the cost of the premiums they pay. Taxpayers who itemize their deductions can deduct the cost of their health care spending that exceeds a threshold (this used to be above 7.5 percent of their adjusted gross income, or AGI, until the new health law raises that amount to above 10 percent of AGI in future years). For more background on the tax treatment of various forms of health care spending, see Joint Committee on Taxation, "Background Materials for Senate Committee on Finance Roundtable on Health Care Financing," JCX-27-09, May 8, 2009.

5. For many decades, the employer sector has provided an effective "civil society" alternative to reliance on government and politics to determine most of the rules and details for private health insurance. Employers (particularly large ones) had both sufficient political clout to resist excessive government regulation and the administrative capacity to deliver health benefits effectively to working Americans.

6. One of the main reasons that most private employers have had a freer hand to adjust their health care coverage and experiment with different types of health care delivery is that the Employee Retirement Income Security Act of 1974 (ERISA) shields self-insured employers from the excesses of state-based health insurance regulation, such as costly mandated benefits and barriers to coordinating benefits across state boundaries. With such federal preemption of state insurance regulation, large employers have been less subject to the political scrutiny typical of public insurance programs, which tend to insist on a one-size-fits-all, centralized set of standards.

7. Spells of unemployment present more difficult economic choices: paying for more expensive COBRA continuation coverage through a former employer's health plan, purchasing individual coverage without tax advantages while subject to health risk underwriting, or preserving diminished economic resources for other purposes. (Standardly referred to by this short abbreviation, the full name of such coverage for former employees stems from the law that authorized it—the Consolidated Omnibus Reconciliation Act of 1985 [COBRA].)

8. Grandfathered employers cannot raise deductibles or overall out-of-pocket limits by more than 15 percent beyond the rate of growth in medical

inflation after March 23, 2010. Similar limits (or $5 above medical inflation, if the latter is more than the maximum percentage increase amount) apply to future increases in copayment amounts. Grandfathered employers also cannot increase a percentage coinsurance cost-sharing provision in their pre–March 23, 2010, health plan by any amount at all. They are further limited in how much they can decrease their contribution rate towards the cost of an employee's coverage (by no more than 5 percent). Moreover, employers wishing to maintain grandfathered protection of their health plan were initially locked into their current insurance carrier (a subsequent reinterpretation of the grandfathering rules allows a change in insurers—as long as the plan's provisions themselves do not also change substantially). U.S. Department of Health and Human Services, "Amendment to Regulation on 'Grandfathered' Health Plan under the Affordable Care Act," 2010, www.hhs.gov/ociio/regulations/grandfather/factsheet.html.

9. Mercer, "Even as Reform Pushes Up Benefits Cost, Employers Will Take Steps to Hold 2011 Increases to 5.9%," September 8, 2010, www.mercer.com/press-releases/1391585.

10. The drafters of the interim regulations for grandfathering issued by HHS in June 2010 estimated that as few as 20 percent of small employers and 36 percent of large employers would remain grandfathered by 2013, depending on various assumptions about future changes in the levels of health plan premiums, cost sharing, and employer contributions. Midrange assumptions by HHS rulemakers were that 34 percent of small employers and 55 percent of large employers would remain grandfathered by 2013. "Fact Sheet: Keeping the Health Plan You Have: The Affordable Care Act and 'Grandfathered' Health Plans," June 14, 2010, www.healthreform.gov/newsroom/keeping_the_health_plan_you_have.html.

11. For example, the PPACA prohibited lifetime health care benefits limits, beginning in September 2010, and it started to require that annual limits on the dollar value of coverage be "reasonable." The new health care law also requires employers to make coverage available to eligible dependent children of covered employees up to age twenty-six (even if they do not qualify as "dependents" for tax purposes or live in another state, or even if they are married and have young children of their own!). The PPACA provision shows modest restraint by stopping short of requiring coverage of grandchildren under a grandparent's insurance policy (in that case, they are not "grandfathered"!). United Healthcare, "Dependent (Adult Child) Coverage to Age 26," United for Reform Resource Center, www.uhc.com/united_for_reform_resource_center/health_reform_provisions/dependent_coverage_to_age_26.htm. Mercer estimates that these early requirements alone pushed up average ESI premiums from a little over 2 percent overall to somewhat larger premium increases for smaller companies. Mercer, "Even as Reform Pushes Up Benefits

Cost, Employers Will Take Steps to Hold 2011 Increases to 5.9%"; see also Bruce Japsen, "Insuring College Students May Add 2 Percent to Medical Premiums," *Chicago Tribune*, July 14, 2010, http://articles.chicagotribune.com/2010-07-14/business/sc-biz-0715-health-students-20100714_1_health-reform-premiums-benefits-consulting-firm-mercer.

12. Kevin Sack, "Health Plan from Obama Spurs Debate," *The New York Times*, July 23, 2008, www.nytimes.com/2008/07/23/us/23health.html.

13. The terms of what such "minimum essential coverage" includes await further definition by regulators. "Affordable" coverage cannot set an employee's share of premiums greater than 9.5 percent of his or her household income.

14. The first requirement was already imposed under a previous federal law, the Health Insurance Portability and Accountability Act (HIPAA) of 1996. The second requirement, imposed to some more limited degree already by HIPAA, would apply to all employer plans (including any grandfathered ones) by 2014.

15. The PPACA requires that, as of 2014, insurance premiums for smaller employer groups (fewer than fifty employees) will be allowed to vary only by coverage tier, number of dependents, geographic region, age (within a 3-to-1 ratio), and tobacco use (within a 1.5-to-1 ratio). Those small-group plans will also be limited as to the maximum amount of cost sharing per enrollee.

16. Dan Diamond, "Golden State Looks On as Others Threaten to Cut Medicaid," November 17, 2010, www.californiahealthline.org/road-to-reform/2010/golden-state-looks-on-as-others-threaten-to-cut-medicaid.aspx.

17. Shawn Tully, "Documents Reveal AT&T, Verizon, Others, Thought About Dropping Employer-Sponsored Benefits," *Fortune*, May 6, 2010, http://money.cnn.com/2010/05/05/news/companies/dropping_benefits.fortune/.

18. The employer mandate penalties are actually designed in a more complex manner than the short description in the main text above. The penalty is the lower of two calculated amounts: $2,000 a year (or its monthly equivalent) for each full-time worker, after excluding the first thirty full-time workers at a firm with fifty or more employees — if an employer does not offer qualified coverage and at least one of the employer's workers receives other subsidized coverage in a health benefits exchange; or $3,000 for each full-time worker at a firm that receives a premium credit through coverage in a state health benefits exchange instead of taking "unaffordable" coverage offered by the employer (see further below in main text).

19. Diana Furchtgott-Roth, "Healthcare's Impact on the Low-Skilled Worker," May 6, 2010, www.realclearmarkets.com/articles/2010/05/06/healthcare_and_low-skilled_workers_98451.html.

20. The PPACA mandates that all businesses begin to report to the Internal Revenue Service the amount and source of all goods they purchase from a vendor after 2011, when the aggregate purchases total over $600 annually. This unprecedented blizzard of paperwork will impose a particularly costly burden

on small businesses—particularly start-up firms—that lack the administrative capacity to handle such recordkeeping obligations.

21. Internal Revenue Service, "National Taxpayer Advocate Submits Mid-Year Report to Congress; Identifies Priority Challenges and Issues for Upcoming Year," IR-2010-83, July 7, 2010, www.irs.gov/newsroom/article/0,,id=225270,00 .html.

22. Center for Health Transformation, "Corporate Costs as Result of Healthcare Reform Law," www.healthtransformation.net/galleries/ wallcharts/Corporate%20Costs%2011.09.10.pdf.

23. National Football League, "'Bill Parcells: Reflections on a Life in Football' Airs on NFLN," November 8, 2010, www.nfl.com/nflnetwork/ story/09000d5d81bf75c9/article/bill-parcells-reflections-on-a-life-in-football-airs-on-nfln.

24. A world in which different workers at the same income level would receive substantially different levels of tax subsidies for health coverage—based solely on whether their employer offered them coverage or did not—and a lucrative new entitlement could be limited to a much smaller number of low- and moderate-income voters is simply not politically sustainable. See, for example, James C. Capretta, "The Senate Health Care Bill's 'Firewall' Creates Disparate Subsidies," December 11, 2009, www.heritage.org/research/ reports/2009/12/the-senate-health-care-bills-firewall-creates-disparate-subsidies; James C. Capretta, "An Entitlement Certain to Grow in Spite of 'Firewalls,'" Kaiser Health News, January 7, 2010, www.kaiserhealthnews.org/ Columns/2010/January/010710capretta.aspx; James C. Capretta, "CBO and a Firewall that Will Never Hold," National Review Online, October 7, 2009, www .nationalreview.com/critical-condition/48164/cbo-and-firewall-will-never-hold/james-c-capretta.

25. Douglas Holtz-Eakin and Cameron Smith, "Labor Markets and Health Care Reform: New Results," American Action Forum, May 2010, www .heartland.org/custom/semod_policybot/pdf/27712.pdf.

26. John Mackey, "The Whole Foods Alternative to ObamaCare," The Wall Street Journal, August 11, 2009, http://online.wsj.com/article/SB1000142405297 0204251404574342170072865070.html.

Impact on. . .Taxpayers

1. "Remarks by the President on Health Insurance Reform in Fairfax, Virginia, George Mason University Patriot Center, Fairfax, Virginia, March 19, 2010," www.whitehouse.gov/the-press-office/remarks-president-health-insurance-reform-fairfax-virginia; "Barack Obama and Joe Biden's Plan to Lower Health Care Costs and Ensure Affordable, Accessible Health Coverage for All," www.barackobama.com/pdf/issues/HealthCareFullPlan.pdf.

2. Peter Orszag, Director of the Office of Management and Budget, "No Gimmick," March 4, 2010, www.whitehouse.gov/omb/blog/10/03/04/No-Gimmick/.

3. Letter to House Speaker Nancy Pelosi from CBO Director Douglas Elmendorf on the cost estimates for the Patient Protection and Affordable Care Act, March 20, 2010, www.cbo.gov/ftpdocs/113xx/doc11379/AmendReconProp.pdf.

4. Ibid.

5. Ibid.

6. Ibid.

7. Ibid.

8. Ibid.

9. "Health Care Reform: An Initial Checkup, With a Keynote Address by Indiana Governor Mitchell E. Daniels Jr.," American Enterprise Institute, June 15, 2010, http://www.aei.org/video/101258.

10. "Remarks by the President to a Joint Session of Congress," The White House, Office of the Press Secretary, September 9, 2009, www.whitehouse.gov/the_press_office/Remarks-by-the-President-to-a-Joint-Session-of-Congress-on-Health-Care/.

11. Senate Budget Committee Republican Staff, "Budget Perspective: The Real Deficit Effect of the Health Bill," December 22, 2009. This estimate was made on the Senate version of ObamaCare, not the final bill. The Senate bill is close to what was passed in final form; therefore, $2.3 trillion is likely to be an accurate ten-year estimate of spending in the final bill as well.

12. See letter to House Speaker Nancy Pelosi from CBO Director Douglas Elmendorf on the cost estimates for the Patient Protection and Affordable Care Act, March 20, 2010, www.cbo.gov/ftpdocs/113xx/doc11379/AmendReconProp.pdf. Chief Medicare Actuary Richard Foster used different assumptions and concluded, "Net Medicare savings are estimated to total $575 billion for fiscal years 2010–2019." Richard S. Foster, "Estimated Financial Effects of the 'Patient Protection and Affordable Care Act,' as Amended," Office of the Actuary, Centers for Medicare & Medicaid Services, April 22, 2010, www.cms.gov/ActuarialStudies/Downloads/PPACA_2010-04-22.pdf.

13. "Weekly Address: Medicare Officially Safer After Health Reform," The White House, August 7, 2010, www.whitehouse.gov/blog/2010/08/07/weekly-address-medicare-officially-safer-after-health-reform.

14. The spending cuts and tax increases in ObamaCare that impact the Medicare Trust Fund total about $350 billion over ten years. If all of these tax increases and spending cuts were dedicated entirely to Medicare and not to a new entitlement, ObamaCare's "surplus" of $124 billion would turn instantly into a more than $200 billion deficit over the next decade.

15. Congressional Budget Office, "Effects of the Patient Protection and

Affordable Care Act on the Federal Budget and the Balance in the Hospital Insurance Trust Fund," December 23, 2009, http://cboblog.cbo.gov/?p=448; and Solomon M. Mussey, Office of the Actuary, Centers for Medicare & Medicaid Services, "Estimated Effects of the 'Patient Protection and Affordable Care Act,' as Amended," on the Year of Exhaustion for the Part A Trust Fund, Part B Premiums, and Part A and Part B Coinsurance Amounts, memorandum, April 22, 2010, www.cms.gov/ActuarialStudies/Downloads/PPACA_Medicare_2010-04-22.pdf.

16. Henry J. Kaiser Family Foundation, "Health Care Reform and the CLASS Act," April 2010, www.kff.org/healthreform/upload/8069.pdf.

17. Lori Montgomery, "Proposed Long-Term Insurance Program Raises Questions," The Washington Post, October 27, 2009, www.washingtonpost.com/wp-dyn/content/article/2009/10/27/AR2009102701417.html.

18. James C. Capretta and Brian Riedl, "The CLASS Act: Repeal Now, or Face Permanent Taxpayer Bailout Later," Backgrounder 2441, Heritage Foundation, July 22, 2010, www.heritage.org/Research/Reports/2010/07/The-CLASS-Act-Repeal-Now-or-Face-Permanent-Taxpayer-Bailout-Later.

19. Robert Pear, "House Passes Bill Averting Cut in Medicare Reimbursements," The New York Times, December 9, 2010, www.nytimes.com/2010/12/10/health/policy/10docs.html?_r=1&emc=tnt&tntemail0=y.

20. Peter Ferrara and Larry Hunter, "How ObamaCare Guts Medicare," The Wall Street Journal, September 9, 2010, http://online.wsj.com/article/SB10001424052748703649004575437311393854940.html; Cathy Ruse and Jerry Birmelin, "ObamaCare's Growing List of Casualties: First Hospitals, Then Politicians," The Daily Caller, December 17, 2010, http://dailycaller.com/2010/12/17/the-growing-list-of-ObamaCares-casualties-first-hospitals-then-politicians/; James A. Bacon, "Casualties Heavy at Hospitals," The Washington Times, August 27, 2010, www.washingtontimes.com/news/2010/aug/27/casualties-heavy-at-hospitals/.

21. Douglas Holtz-Eakin and James C. Capretta, "Resetting the 'ObamaCare' Baseline," December 16, 2010, www.politico.com/news/stories/1210/46428.html.

22. Congress regularly does this now to protect the middle class from the alternative minimum tax.

23. Centers for Medicare & Medicaid Services, "2010 Poverty Guidelines," www.cms.gov/MedicaidEligibility/downloads/POV10Combo.pdf.

24. Letter to Senator Evan Bayh from CBO Director Douglas Elmendorf on the premium effects of the Patient Protection and Affordable Care Act, November 30, 2009, www.cbo.gov/ftpdocs/107xx/doc10781/11-30-Premiums.pdf.

25. Kaiser Family Foundation, "Summary of Health Reform Law," March 26, 2010, www.kff.org/healthreform/upload/8061.pdf.

26. Census Bureau, "POV01: Age and Sex of All People, Family Members

and Unrelated Individuals Iterated by Income-to-Poverty Ratio and Race: 2009," www.census.gov/hhes/www/cpstables/032010/pov/new01_400_01.htm.

27. Letter to House Speaker Nancy Pelosi from CBO Director Douglas Elmendorf on the cost estimates for the Patient Protection and Affordable Care Act, March 20, 2010, www.cbo.gov/ftpdocs/113xx/doc11379/AmendReconProp.pdf.

28. Stephanie Rennane and C. Eugene Steuerle, "Health Reform: A Two-Subsidy System," Urban Institute, April 2, 2010, www.taxpolicycenter.org/numbers/Content/PDF/S10-0001.pdf.

29. Douglas Holtz-Eakin and Cameron Smith, "Labor Markets and Health Care Reform: New Results," American Action Forum, www.heartland.org/custom/semod_policybot/pdf/27712.pdf, May 2010.

30. Philip Bredesen, "ObamaCare's Incentive to Drop Insurance," *The Wall Street Journal*, October 21, 2010, http://online.wsj.com/article/SB10001424052702304510704575562643804015252.html.

31. Congressional Budget Office, "Monthly Budget Review," November 5, 2010, www.cbo.gov/ftpdocs/118xx/doc11873/NovemberMBR.pdf.

32. Congressional Budget Office, *An Analysis of the President's Budgetary Proposals for Fiscal Year 2011*, March 2010, www.cbo.gov/ftpdocs/112xx/doc11280/03-24-APB.pdf.

Impact on. . .You and Your Constitutional Rights

1. George F. Will, "A Battle Won, but a Victory?," *The Washington Post*, March 23, 2010, www.washingtonpost.com/wp-dyn/content/article/2010/03/22/AR2010032201528.html.

2. Ibid.

3. Thomas P. Miller, "Health Reform: Only a Cease-Fire in a Political Hundred Years' War," *Health Affairs* 29, no. 6 (June 2010), www.aei.org/docLib/MillerHealthAffairsPoliticsofHealthReform.pdf.

4. Ibid.

5. Curtis Copeland, "Initial Final Rules Implementing the Patient Protection and Affordable Care Act," Congressional Research Service, December 10, 2010, http://healthreform.kff.org/~/media/Files/KHS/docfinder/crs18regulations.pdf.

6. Ibid. The Administrative Conference of the United States has also said that "agencies are perceived by commenters as more likely to accept changes in a rule that has not been promulgated as a final rule—and potential commenters are more likely to file comments in advance of the agency's 'final' determination." "Recommendation 95-4, Procedures for Noncontroversial and Expedited Rulemaking," www.law.fsu.edu/library/admin/acus/305954.html.

7. "Fact Sheet: Keeping the Health Plan You Have: The Affordable Care

Act and 'Grandfathered' Health Plans," June 14, 2010, www.healthreform.gov/ newsroom/keeping_the_health_plan_you_have.html.

8. Jay Heflin, "IRS Delays Cost Report on Healthcare Reform," October 12, 2010, http://thehill.com/blogs/on-the-money/domestic-taxes/123811-irs-delays-cost-report-on-health-care. Section 9002 of the PPACA specifically requires that the inclusion of the aggregate cost of employer-sponsored health coverage on employees' W-2 tax information forms "shall apply to taxable years beginning after December 31, 2010."

9. Janet Adamy, "McDonald's May Drop Health Plan," *The Wall Street Journal*, October 30, 2010, http://online.wsj.com/article/SB10001424052748703 431604575522413101063070.html.

10. Joseph Antos, "Long May She Waive," *The American*, December 14, 2010, www.american.com/archive/2010/december/long-may-she-waive.

11. "Obama Opposed Mandated Insurance During Campaign," www .youtube.com/watch?v=EwzYVEunPQ0&feature=related.

12. Interview on CNN's American Morning, February 5, 2008, http:// transcripts.cnn.com/TRANSCRIPTS/0802/05/ltm.02.html.

13. Jesse Lee, "The President Spells Out His Vision on Health Care Reform," June 3, 2009, www.whitehouse.gov/blog/The-President-Spells-Out-His-Vision-on-Health-Care-Reform.

14. Thomas P. Miller, "False Hopes, Empty Illusions," *USA Today*, June 10, 2009; Thomas P. Miller, "The Last Detail(s)," American Enterprise Institute, June 4, 2009, www.aei.org/speech/100065; Thomas P. Miller, "Healthcare Dreams, Healthcare Realities," *The American*, July 16, 2009, www.american .com/archive/2009/july/healthcare-dreams-healthcare-realities.

15. Randy Barnett, "A Noxious Commandment," *The New York Times*, December 13, 2010, www.nytimes.com/roomfordebate/2010/12/13/a-fatal-blow-to-obamas-health-care-law/an-unconstitutional-commandment.

16. *Commonwealth of Virginia ex rel. Cuccinellil v. Sebelius*, 720 F. Supp. 2d 598 (E.D. Va. 2010); *State of Florida, et al. v. United States Department of Health and Human Services, et al.*, 645 So. 2d 578 (N.D. Fla. 2010).

17. *Wickard v. Filburn*, 317 U.S. 111 (1942).

18. *Gonzalez v. Raich*, 545 U.S. 1 (2005).

19. U.S. District Court Judge Henry E. Hudson, Memorandum Opinion, Civil Action No.3:10CV188-HEH, December 13, 2010, www.oag.state.va.us/ PRESS_RELEASES/Cuccinelli/Health%20Care%20Memorandum%20 Opinion.pdf.

20. Judge George Caram Steeh, Order, *Thomas More Law Center, et al. v. Barack Hussein Obama*, 720 F. Supp. 2d 882 (E.D. Mich. 2010); Judge Norman K. Moon, Memorandum Opinion, *Liberty University, Inc. et al. v. Timothy Geithner, Kathleen Sebelius, Hilda Solis, and Eric Holder*, 106 A.F.T.R. 2d (RIA) 7174 (W.D. Va. 2010).

21. *United States v. Morrison,* 529 U.S. 598 (2000) at 614.

22. In *Perez,* the Court upheld a federal prohibition on extortionate credit transactions, but in *Lopez,* it found that making it a federal offense for any individual knowingly to possess a firearm in a school zone *exceeded* the Commerce Clause authority of Congress. Similarly, the Court concluded in *Morrison* that civil remedies imposed under the Violence Against Women Act involved only noneconomic, violent criminal conduct (gender-motivated violence), did not substantially affect interstate commerce, and therefore lacked constitutional authority.

23. Judge Roger Vinson, Order and Memorandum Opinion, *State of Florida, et al. v. United States Department of Health and Human Services, et al.,* 645 So. 2d 578 (N.D. Fla. 2010).

24. Ibid. at 28.

25. Ibid. at 60–61.

26. The Congressional Research Service, which provides Congress with analysis on the constitutionality of pending legislation, had advised Congress on July 24, 2009, that "it is a novel issue whether Congress may use this [Commerce] clause to require an individual to purchase a good or service." Jennifer Staman and Cynthia Brougher, "Requiring Individuals to Obtain Health Insurance: A Constitutional Analysis," http://assets.opencrs.com/rpts/R40725_20090724.pdf, p. 3.

27. Vinson, Order and Memorandum Opinion at 63.

28. N. C. Aizenman, "Florida Judge Considers 20-State Challenge of Health-Care Law," *The Washington Post,* December 17, 2010, A11.

29. The Ninth Amendment to the U.S. Constitution says, "The enumeration in the Constitution of certain rights shall not be construed to deny or disparage others retained by the people." The Tenth Amendment says, "The powers not delegated to the United States by the Constitution, or prohibited by it to the States, are reserved to the States respectively, or to the people."

30. The average federal matching assistance percentage is 57 percent, and some states with particularly low-income populations receive more than three out of four dollars in total Medicaid costs courtesy of out-of-state taxpayers.

31. In the reconciliation bill (Public Law 111–152), Congress modified the eligibility standard from 133 percent to 138 percent of the federal poverty level by "disregarding" additional income of up to 5 percent of FPL.

32. For example, the Texas Health and Human Services Commission for the House Select Commission on Federal Legislation estimated in April 2010 that the new health care law will increase health care costs for the state of Texas by $27 billion from 2014 to 2023. The state of Indiana estimated that the full impact of the PPACA on its budget would amount to an additional $3.6 billion from 2011 to 2020. Evelyne Baumrucker and Bernadette Fernandez, Congressional Research Service, "Variation in Analyses of PPACA's Fiscal Impact on States,"

memorandum, September 8, 2010, http://healthreform.kff.org/~/media/
Files/KHS/Scan/CRS%20State%20Impact%20of%20PPACA.pdf.

33. Miller, "Health Reform: Only a Cease-Fire in a Political Hundred Years'
War," at 1102.

34. *South Dakota v. Dole*, 483 U.S. 203 (1987) at 207–10.

35. James F. Blumstein, "Who's in Charge? More Legal Challenges to
the Patient Protection and Affordable Care Act," presentation at American
Enterprise Institute, December 6, 2010, www.aei.org/event/100342.

36. Bill McCollum, "A Failed Public Policy," *The Washington Post*, December 17,
2010, www.washingtonpost.com/wp-dyn/content/article/2010/12/16/AR2010
121606734.html.

37. Judge Roger Vinson, Order Granting Summary Judgment, *State of
Florida, et al. v. United States Department of Health and Human Services, et al.* (N.D.
Fla. 2011), www.flnd.uscourts.gov/announcements/documents/10cv91doc150
.pdf.

38. Vinson, Order and Memorandum Opinion at 49–50.

39. The Patient Protection and Affordable Care Act, Public Law 111–148,
Section 1311.

40. Thomas M. Christina, "Who's in Charge? More Legal Challenges to the
Patient Protection and Affordable Care Act," December 6, 2010, www.aei.org/
video/101357.

41. Ted Cruz and Mario Loyola, "Reclaiming the Constitution: Towards
an Agenda for State Action," Texas Public Policy Foundation, November 2010,
www.texaspolicy.com/pdf/2010-11-RR11-TenthAmendment-ml.pdf.

42. *Printz v. United States*, 521 U.S. 898 (1997) at 930.

43. James Taranto, "Stealth Socialism," December 17, 2010, http://online
.wsj.com/article/SB10001424052748704034804576025731120969862.html.

What We Should Do Instead to Get Reform Right

1. The American people have made it very clear that they are opposed to
ObamaCare. Support for repeal, according to Rasmussen Reports, has ranged
from 50 to 63 percent of voters since the law was enacted on March 23, 2010.
Most think the law is bad for the country, will increase health care costs,
and increase the federal deficit. The latest available average of the ten most
recent national polls on the health care law, as compiled by *Real Clear Politics*
in January 2011, indicated that 47.5 percent opposed it and only 41 percent
favored it. When voters opposed to the law were asked why they opposed it,
the pollster Zogby International found that the main reason was that it would
give the government "too much control" over their health care decisions.
Rasmussen Reports, "Support for Health Care Repeal at 60%," December 27,

2010,   www.rasmussenreports.com/public_content/politics/current_events/
healthcare/december_2010/support_for_health_care_repeal_at_60;   Zogby
International, "51% of Voters Oppose Healthcare Reform Act," April 21, 2010,
www.zogby.com/news/ReadNews.cfm?ID=1849; *Real Clear Politics*, "Obama
and Democrats' Health Care Plan," January 20, 2011, www.realclearpolitics.
com/epolls/other/obama_and_democrats_health_care_plan-1130.html.

2. All of the authors and our colleagues in the market-based health policy
community have written extensively about our ideas for patient-centered health
reform. You will find addresses for our websites with much more information at
the end of the chapter on "What you can do to put the brakes on ObamaCare."

3. Grace-Marie Arnett, ed., *Empowering Health Care Consumers through Tax
Reform*, (Ann Arbor: University of Michigan Press, 1999), www.galen.org/
component,8/action,show_content/id,7/news_id,2205/type,33/.

4. Grace-Marie Turner, "Competition in the Health Care Marketplace,"
testimony before the U.S. Senate Committee on Commerce, Science, and
Transportation, July 16, 2009, www.galen.org/component,8/action,show_
content/id,13/blog_id,1230/type,33/.

5. Mitch Daniels, "Hoosiers and Health Savings Accounts," *The Wall Street
Journal*, March 1, 2010, http://online.wsj.com/article/SB10001424052748704231
30457509160047029066.html.

6. In addition, neither employer-paid insurance premiums nor health care
benefits financed directly by self-insured employers are counted as employee
wages for purposes of the payroll taxes that finance Social Security and Medicare.

7. Purchases of some other forms of private health insurance do receive
more limited tax benefits. Self-employed workers can claim an income tax
deduction for the cost of the premiums they pay. Taxpayers who itemize their
deductions can deduct the cost of their health spending that exceeds a threshold
(above 7.5 percent of their adjusted gross income, or AGI, under previous rules –
until the new health law raises that amount to above 10 percent of AGI in future
years). For more background on the tax treatment of various forms of health
care spending, see Joint Committee on Taxation, "Background Materials for
Senate Committee on Finance Roundtable on Health Care Financing," JCX-27-
09, May 8, 2009, www.jct.gov/publications.html?func=showdown&id=3557.

8. A couple filing a joint income tax return with total wages of $75,000 fell
into the 25 percent marginal tax bracket in 2010. In addition, the combined
employer-employee payroll tax rate was 15.3 percent. Overall, then, the federal
tax break equaled the total federal taxes not paid on $13,000, or (.25 + .153) ×
$13,000 = $5,239.

9. Kaiser Family Foundation and Health Research and Educational Trust,
Employer Health Benefits, 2009 Annual Survey, http://ehbs.kff.org/pdf/2009/7936.
pdf.

10. This credit might initially be the same amount of money for everyone,
before making necessary adjustments to be more generous to people with low

incomes and much higher health care costs. It could be paid out as a "voucher" to eligible individuals or advanced directly to the insurer offering the health plan they choose. The credit would be "refundable" for those individuals who do not pay enough, or any, federal taxes for the tax credit to "offset."

11. Grace-Marie Turner, "Providing Coverage for All Through Private Health Insurance," Galen Institute, December 8, 2008, www.galen.org/component,8/action,show_content/id,13/blog_id,1132/type,33/.

12. Presentation by Richard S. Foster, Chief Actuary, Centers for Medicare & Medicaid Services, at the American Enterprise Institute, May 13, 2009, www.aei.org/docLib/Foster%20AEI2009.pdf.

13. Joseph Antos and Thomas P. Miller, *A Better Prescription: AEI Scholars on Realistic Health Reform*, American Enterprise Institute, 2010.

14. Representative Paul D. Ryan, "A Roadmap for America's Future: Version 2.0," Committee on the Budget, U.S. House of Representatives, January 2010, www.roadmap.republicans.budget.house.gov/UploadedFiles/Roadmap2Final2.pdf.

15. For a discussion of the health policy initiatives in the Roadmap, see Robert E. Moffit and Kathryn Nix, "The Future of Health Care Reform: Paul Ryan's 'Roadmap' and Its Critics," Heritage Foundation Backgrounder no. 2495, December 3, 2010, http://report.heritage.org/bg2495.

16. The tax credits for individuals and their families would be risk-adjusted and income-related, however, and their value would be indexed to a blend of future general price inflation and medical price inflation.

17. James C. Capretta and Tom Miller, "How to Cover Pre-Existing Conditions," *National Affairs*, Summer 2010, www.nationalaffairs.com/publications/detail/how-to-cover-pre-existing-conditions; Thomas Miller and James C. Capretta, "Changing the Name – But Not the Political Game," *Health Affairs Blog*, July 30, 2010, http://healthaffairs.org/blog/2010/07/30/changing-the-name-but-not-the-political-game/.

18. Amy Goldstein, "Health Plans for High-Risk Patients Attracting Fewer, Costing More Than Expected," *The Washington Post*, December 27, 2010, www.washingtonpost.com/wp-dyn/content/article/2010/12/27/AR2010122702343.html.

19. Capretta and Miller, "How to Cover Pre-Existing Conditions."

What You Can Do to Put the Brakes on ObamaCare

1. See letter to Senator Evan Bayh from CBO Director Douglas Elmendorf on the premium effects of the Patient Protection and Affordable Care Act, November 30, 2009, www.cbo.gov/ftpdocs/107xx/doc10781/11-30-Premiums.pdf.

2. Ricardo Alonso-Zaldivar, "Fact Check: White House Health Savings Challenged," September 13, 2010, www.boston.com/business/healthcare/

articles/2010/09/13/fact_check_white_house_health_savings_challenged/.

3. Richard S. Foster, "Estimated Financial Effects of the 'Patient Protection and Affordable Care Act,' as Amended, Office of the Actuary," Centers for Medicare & Medicaid Services, April 22, 2010, www.cms.gov/ActuarialStudies/Downloads/PPACA_2010-04-22.pdf.

4. Ibid. (The Congressional Budget Office puts the ten-year saving at $136 billion. Congressional Budget Office cost estimate for the PPACA, as amended by H.R. 4872, March 20, 2010, www.cbo.gov/ftpdocs/113xx/doc11379/AmendReconProp.pdf.)

5. Letter to Senator Charles E. Grassley from CBO Director Douglas Elmendorf, October 8, 2010, http://op.bna.com/hl.nsf/id/bbrk-8a7t97/$File/ActuaryCMSOct2010.pdf.

6. Douglas Holtz-Eakin and James C. Capretta, "Resetting the 'ObamaCare' baseline," December 16, 2010, www.politico.com/news/stories/1210/46428.html.

7. See letter to House Speaker Nancy Pelosi from CBO Director Douglas Elmendorf on the cost estimates for the Patient Protection and Affordable Care Act, March 20, 2010, www.cbo.gov/ftpdocs/113xx/doc11379/AmendReconProp.pdf; Richard S. Foster, "Estimated Financial Effects."

8. John Reichard, "Analyst: Health Law Means Upheaval for Health Care Economy," CQ HealthBeat News, November 11, 2010, www.cq.com/doc/hbnews-3761576?wr=RDlYTlRja3lSajRoWWFoZGNwN084UQ.

9. John D. Shatto and M. Kent Clemens, Office of the Actuary, Centers for Medicare & Medicaid Services, "Projected Medicare Expenditures under an Illustrative Scenario with Alternative Payment Updates to Medicare Providers," memorandum, August 5, 2010, www.cms.gov/ReportsTrustFunds/downloads/2010TRAlternativeScenario.pdf, p. 6.

10. "Joining the ObamaCare Suit," The Wall Street Journal, November 18, 2010, http://online.wsj.com/article/SB1000142405274870439360457561510413 5953066.html?mod=djemEditorialPage_h.

About the Authors

The authors of *Why ObamaCare Is Wrong for America* have been working together for years to offer a better prescription for health reform—one that puts doctors and patients in control of health care decisions. They have been deeply involved in following the debate over health reform and educating the American people about the impact of policy initiatives. They have spoken across the country, done thousands of radio and television interviews, and written hundreds and hundreds of commentaries, blog posts, and papers explaining the legislation and its impact on America. They have seen and heard the outrage in America over Obama-Care and are on the side of Americans who strongly oppose this law. They work for four different conservative think tanks that have led the fight to educate the American people about the impact of ObamaCare. All are known as top experts on health care reform and ObamaCare in particular and are uniquely able to tell the story of why ObamaCare is so unpopular and what we need to do instead.

Grace-Marie Turner is the president of the Galen Institute, a public policy research organization she founded in 1995 to promote an informed debate over free-market ideas for health care reform. She speaks and writes extensively about incentives to promote a more competitive, patient-centered marketplace in the health care sector. She testifies regularly before Congress and advises senior government officials, governors, and state legislators on health care policy. Grace-Marie served a three-year term as a member of the National Advisory Council for the Agency for Healthcare Research and Quality and served as a member of the Medicaid Commission, charged with making recommendations to modernize and improve Medicaid. She is a founder and facilitator of the Health Policy Consensus Group, which serves as a forum for analysts from market-oriented think tanks around the country to analyze and develop policy recommendations. She is the editor of *Empowering Health Care Consumers Through Tax Reform* and produces a widely read weekly electronic newsletter, *Health Policy Matters*. She has been published in major newspapers, including *The Wall Street Journal* and *USA Today*, and has appeared on ABC's *20/20* and on hundreds of radio and television programs in the United States. She received the 2007 Outstanding Achievement Award for Promotion of Consumer Driven Health Care from Consumer Health World. Her early career was in politics and journalism, where she received numerous awards for her writing on politics and economics. She has a BA in journalism from the University of New Mexico.

James C. Capretta, a Fellow at the Ethics and Public Policy Center (EPPC) and Project Director of ObamaCareWatch.org, was an associate director at the White House Office of Management and Budget (OMB) from 2001 to 2004, where he was the top budget official for health care, Social Security, education, and welfare programs. At EPPC, Mr. Capretta studies and provides commentary on a wide range of public policy and economic issues, with a focus on health care and entitlement reform, U.S. fiscal policy, and global population aging. His essays and articles have appeared in numerous print and online publications, including *USA Today, Politico, Health Affairs, National Affairs, Kaiser Health News, The Weekly Standard*, and *Tax Notes*, among others. He is the author of the blog *Diagnosis* and is a frequent contributor to *National Review Online*. Mr. Capretta has also testified before Congress and appeared as a commentator on *PBS NewsHour*, FOX News, FOX Business News, CNBC, MSNBC, EWTN, and numerous national and local radio programs. Earlier in his career, Mr. Capretta served for a decade in Congress as a senior analyst for health care issues and for three years as a budget examiner at OMB. He has an MA in public policy studies from Duke University, and he graduated from the University of Notre Dame in 1985 with a BA in government.

Thomas P. Miller is a resident fellow at the American Enterprise Institute, where he focuses on health care policy, with particular emphasis on such issues as information transparency, health insurance regulation, and consumer-driven health care. He was a member of the National Advisory Council for the Agency for Healthcare Research and Quality from 2007 to 2009. Before joining AEI, Mr. Miller served for three years as a senior health economist for the Joint Economic Committee, where he organized a series of hearings focusing on promising reforms in private health care markets. He has also been a director of health policy studies at the Cato Institute and director of economic policy studies at the Competitive Enterprise Institute. Before coming to Washington, he had a real life as a trial attorney, journalist, and radio broadcaster (including several seasons as the play-by-play voice of the Davidson College Wildcats basketball team). Tom Miller's writing has appeared in such publications as *Health Affairs, The Wall Street Journal, The New York Times, The Washington Post, Los Angeles Times, USA Today, Reader's Digest, National Review, The American, Law and Contemporary Problems, Regulation,* and *Cato Journal*. He has testified before various congressional committees on issues including the uninsured, Medicare prescription drug benefits, medical savings accounts, health insurance tax credits, genetic information, Social Security, federal reinsurance of catastrophic risks, and terrorism insurance. Miller holds a bachelor's degree in political science from New York University and a law degree from Duke University.

Robert Emmet Moffit, a seasoned veteran of more than three decades in Washington policy making, is a senior fellow at The Heritage Foundation's Center for Policy Innovation. Moffit specializes in health care and entitlement programs, including Medicare. He was a senior official of the U.S. Department of Health and Human Services (HHS) and the Office of Personnel Management (OPM) during the Reagan administration. He directed Heritage's Center for Health Policy Studies from 2003 until June 2010. He was one of only a few conservatives to make *Modern Healthcare* magazine's August 2010 list of "The 100 Most Powerful People in Healthcare." He has appeared on the major cable news channels as well as the broadcast networks and is quoted regularly by *USA Today* and other leading newspapers. His analysis and commentary have been cited or published by *The New York Times, The Wall Street Journal, New York Post,* and *The Washington Post,* among scores of other newspapers and websites. Moffit has also been published in professional journals, such as *Health Affairs, Health Systems Review, Harvard Health Policy Review, Journal of Contemporary Health Law and Policy, Postgraduate Medicine,* and *Journal of Medicine and Philosophy*. Moffit holds both a master's degree and a doctorate in political science from the University of Arizona and received his bachelor's degree in political science from LaSalle University in Philadelphia.